Origins of The Organic Agriculture Debate

Origins of
The Organic Agriculture Debate

Thomas R. DeGregori

Iowa State Press
A Blackwell Publishing Company

THOMAS R. DEGREGORI, PH.D., is a professor of economics at the University of Houston, Texas, and author of numerous scholarly books, articles, and reviews. His fields of expertise are economic development; technology and science in economic development; and African, Asian, and Caribbean economic development. Dr. DeGregori has served on many editorial boards and boards of directors and is currently on the Board of Directors of the American Council on Science and Health. He is a popular speaker, lecturer, and consultant both nationally and internationally.

© 2004 Iowa State Press
A Blackwell Publishing Company
All rights reserved

Iowa State Press
2121 State Avenue, Ames, Iowa 50014

Orders: 1-800-862-6657
Office: 1-515-292-0140
Fax: 1-515-292-3348
Web site: www.iowastatepress.com

Authorization to photocopy items for internal or personal use, or the internal or personal use of specific clients, is granted by Iowa State Press, provided that the base fee of $.10 per copy is paid directly to the Copyright Clearance Center, 222 Rosewood Drive, Danvers, MA 01923. For those organizations that have been granted a photocopy license by CCC, a separate system of payments has been arranged. The fee code for users of the Transactional Reporting Service is 0-8138-0513-9/2004 $.10.

⊗ Printed on acid-free paper in the United States of America

Library of Congress Cataloging-in-Publication Data

DeGregori, Thomas R.
    Origins of the organic agriculture debate \ Thomas R. DeGregori.
      p. cm.
    ISBN 0-8138-0513-9 (alk. paper)
    1. Organic farming.   I. Title.
    S605.5.D44 2003
    631.5'84—dc21   2003014331

The last digit is the print number: 9 8 7 6 5 4 3 2 1

Contents

Preface, vii

Introduction, xv

1.   Science, Technology, and the Critics of Modernity   3

2.   Science, Integrated Inquiry, and Verification   9

3.   Reductionism: Sin, Salvation, or Neither?   21

4.   On the Trail of DNA: Genes and Heredity   27

5.   Vitalism and Homeopathy   41

6.   Disenchantment and the Cost of Rejected Knowledge   53

7.   Rejected Knowledge, Nature, and the Occult   65

8.   Vitalism, the Organic, and the Precautionary Principle   83

9.   Feeding Six Billion People   95

10.   Romantics and Reactionaries   133

11.   Risk, Representation, and Change   151

Epilogue: Science, Technology, and Humanity   161

References   169

Index   205

Dedication

To
Michael, Luke, and Alejandro
(in birth order)
and their siblings and cousins yet to be born

Preface

This Book

A boundless sense of wonder and curiosity has led humans to ask many questions about why and how and what next. It is out of this spirit of questioning that the active, problem-solving human mind has expanded the scope of human understanding, created science and technology, and in the process made a better life for all of us. This book focuses on a particular aspect of this process in following a track of scientific inquiry, primarily in chemistry and biology, from Lavoisier to the present, in which humans have explored the fundamental elements of living processes right on down to the nucleic acids that constitute DNA. This advancing knowledge has led to dramatic reductions in disease and death, provided better food and nutrition for a growing population, and expanded and bettered all aspects of human life. I argue here and elsewhere that advancing knowledge is a resource-creating process that underlies my conviction, for which there is more than ample historical and theoretical support, that the bettering process of the human endeavor is open-ended and can continue through time.

Despite the obvious improvements that this science has wrought—the statistics that I give here and elsewhere are astounding—many are adamantly opposed to this scientific inquiry, calling it reductionist. This opposition, often based on irrational fear, is as old as the science that it counters and I follow its development through the nineteenth and twentieth centuries in a kind of a double helix as I contrast advances in science, medicine, and agriculture with the oppositional beliefs—

homeopathy and "organic" agriculture—that continue to the present. I find the thread of continuity that runs through these various antiscience views to be a belief in an unmeasurable, essentially unknowable vital force, or vitalism. This is a partisan book in that I argue that these vitalist beliefs are largely harmful in their impact.

The vision that I offer of science is larger than the mere statistics of human well-being, however heartening improving human health and life extension may be. Science offers the possibility to be a transcultural unifying force in a diverse world. Critics may point to its shortcomings, which are many as is the case for any human endeavor, but science offers a hope of overcoming the barriers that have historically divided us. It is traditional knowledge, which many are now touting, that defined the differences that allowed some to believe that others were inferior to them and could therefore be treated accordingly.

This Book as Part of a Larger Inquiry

This book is the last of a quadrilogy—*A Theory of Technology* (1985), *Agriculture and Modern Technology* (2001), and *The Environment, Our Natural Resources and Modern Technology* (2002)— all of which were published by Iowa State University Press (now Iowa State Press, A Blackwell Publishing Company). I have posted on my webpage (www.uh.edu/~trdegreg) a supplementary bibliography for the latest books to keep readers current on important issues that I have discussed, particularly the more controversial ones. The integrating thesis of all of my writing is that modern science and technology have provided an increasing number of us a quality of life and the longevity to enjoy it unprecedented in human history. It also gives us the obligation and opportunity as never before, to use this science and technology to create a better world for all. This does not deny that science and technology have also created weapons of mass destruction, but the data are clear and overwhelming that science and technology have saved vastly more lives than they have taken. We need to understand better the forces of scientific and technological change if we are to control the negative elements of these forces, continue to advance the development of science and technology, and facilitate fuller participation in the benefits of our advancing capability to further the human endeavor.

The thesis that I have been advancing has met opposition across the political spectrum. To an increasing number, the more that science and technology improve our lives, the more fervently they believe that it is harming us. They seek refuge from modernity in "alternative medicine" and "organic" food products. One study points to "one inescapable conclusion: Life on Earth is killing us" (CNS 1998). In this book as in my earlier ones, I ask this question: if science and technology are killing us, why are we living so long? If our food is so lacking in nutrients and our medicine and pharmaceuticals so ineffective, then why are we so healthy? Once again, I expect to receive a deafening silence back. If some who have read my earlier books and articles are tired of reading or hearing my question, believe me, I am also tired of asking it and would appreciate someone making a modest effort to answer it.

In the epilogue, I make a claim that merits being stated here also. Never in my life has science and the scientist been so overwhelmingly in support of technology as is the case with biotechnology in agriculture and a range of other technologies in food production and human health. Never has the opposition been so organized and the media and public so effectively misled on these issues. Clearly much remains to be done in public science education. Being an optimist, I write books and articles with the uncompromising and undiminished faith that the light of reason will shine through the darkness of even the most organized ignorance, and that science, technology, and other human knowledge and understanding will show us the way to that future that we all desire and that the least privileged individuals desperately need.

When I entered college, the humanism of the Renaissance had an honored place in academia. Today, being a humanist subjects one to attack from the religious radical right with the pejorative, secular humanist. Equal scorn comes from the radical postmodernists, ecofeminists, and deep ecologists who view humanism as speciesist as they prefer a more earth-centric or biocentric view. Once again, if charged, I plead guilty. To me, without a core set of humanistic values, all values about other life-forms and the earth are meaningless. In my judgment, the humanistic values implicit in science and technology are more than capable of creating an intelligent operational philosophy in which the human life process sustains itself in a manner appreciative of the virtue of other forms of life and the beauty of the world, both natural and that made by humans.

My Debt to David Hamilton

I entered college from an upper-middle-class family participating in the wealth and freedom of the richest country the world had ever known. Today, many countries have far surpassed the level of wealth of my youth, which is a very important part of what this book and my life work have been about. I entered college fiercely determined to defend that wealth and freedom. A later generation would be taught that affluence was evil. I was lucky to have teachers like David Hamilton who agreed that affluence and freedom were to be appreciated but that a humanistic belief in the worth and dignity of other humans required that we protect our wealth in the most just and effective way possible by creating the conditions where all have the same opportunity to participate in this enjoyment.

From Hamilton I learned the virtue and value of incremental change and the importance of compromise in a democratic society. Principles are to be put into practice. Politics was not about creating utopias but about formulating policies that improved the lives of the nation's citizens. No one would deny that there are some principles so paramount that one must lose now so better to fight for them tomorrow. But too often today, this is used as an excuse by self-indulgent elitists who seem willing to forsake everything, including the betterment of the less fortunate who would benefit from compromise, to preserve their sense of being pure in their pursuit of a principle. Principles and goals realized a step at a time are no less worthy of being pursued and no less important to their beneficiaries.

Being a development economist by profession and inclination, I learned the virtue of Hamilton's incremental change, be it a larger crop for a previously subsistence farm family, a single light in the house and a spigot drawing clean water just outside it, or off-farm employment for a couple of days each week to earn school fees for the children and a few essential items of consumption. I recall a village in Rajasthan, India, with a single well as the source of water for all household uses. Women and children used to line up before dawn and wait for hours to haul up a few buckets of water from the deep well. A small diesel-powered pump and a large water storage tank with three spigots dramatically changed lives. Drawing water was no longer as laborious, there was always a spigot available without waiting, and the children were in school. And it gave hope for more change to come. Improved nutrition and health, new skills and opportunities, and, most important, children getting an education laid the foundation for continued change. As an enthusiast for high-tech and frontier technologies, then and now, I have also seen the virtues of the incremental addition of basic existing technologies. It has been my privilege to have witnessed the cumulation of these incremental changes combining with other technological changes to bring about transformations in Asia that are unprecedented in human history. These events more than verified what Hamilton had taught and my indebtedness to him.

In recent decades as the doubts about the benefits of science and technology have grown in some segments of society and become established dogma in some areas of academia, radicalized youth have taken to the streets in opposition to science and technology and to the institutions that they identify with science and technology in the mistaken belief that they are defending the poor and powerless of the world. However they may claim solidarity with the downtrodden, there is no evidence that those upon whose behalf they presume to speak wish them to do so. It is obvious to most everyone but the protesters that the poor need better agronomy in agriculture—improved seeds biotech or otherwise, fertilizer, and so on—and the benefits that improvements in technology can bring to all sectors of society. The idealism of some of the protesters may be commendable but when it is informed by "rejected knowledge," great harm can result with those most harmed being those most in need and least able to promote their own needs and aspirations.

Like the postmodernists whom I criticize, I recognize that we all have our biases. However, I believe that free, open, or transparent inquiry is capable through time of sorting out different biases, separating fact from fiction and thereby expanding knowledge and human capability. The narrative that I have been relating and the story that follows is one that describes a human journey in which we, its participants, have been expanding our numbers and an increasing proportion of us are living longer, healthier lives. If we are to continue on this pathway, then we must seek to understand the forces that have brought about this change in the past and are operating today.

I make no apologies for the often assertive tenor of this book. I feel compelled to make strong forthright arguments in favor of a set of ideas and practices as well as set forth strong arguments against what I believe to be wrong ideas and practices. Serious issues require serious debate and no issue is more important than how we will feed nine billion people in less than a half century from now. However, being admittedly assertive is not a license for invectives, name calling, and character assassination. This does not mean that one cannot occasionally make a critical assessment of an individual as long as it is in terms of the ideas the individual expresses. If there is a combative tone, it is over the clash of ideas and not personalities. In this spirit, I welcome strong arguments against the ideas expressed here (or in any of my previous work) as long as those of us engaged in this discourse can do so without maligning the character and impugning the motives and integrity of those with whom they disagree. The issues that we are discussing have become heated, and restraint against personal attacks has not been the order of the day.

One can spend a lifetime like the fabled Midas, obsessed with the need to protect one's wealth and freedom from those who would take it, or one can recognize the potential of science and technology to open the possibility of a better world for all. When I was in graduate school at the University of Texas, the story was told, possibly apocryphal, about the populist professor who was investigated for subversive beliefs before the Texas legislature. When asked whether he believed in private property, he is alleged to have responded, yes, he did and that he believed in it so much that he wanted everyone to have some of it. If I may paraphrase him, I believe so much in the affluence and freedom made possible by science and technology that I want everyone to have the opportunity to have some of it. The working out of this belief is what this book is about.

A Note on Sources and References

Most of the quality scholarly journals have webpages where each issue is posted, sometimes going online even before regular subscribers have received their hard copy. University libraries are acquiring subscriptions that allow their faculty free access to journals that otherwise charge for it. An increasing numbers of us will search for far more articles on the internet than in the library. For faculty at colleges and universities whose library holdings were limited for any number of reasons, the internet has provided opportunity for them to participate more fully in the latest advances in their field. This is particularly true for faculty and libraries in poorer countries, as many subscription-based journals allow free access from addresses in poorer countries.

Libraries, such as that of my own university, that used to plead with faculty to give their journal collections to the library so that they could fill in missing issues and trade or sell the others, now will not even accept them as gifts.

The vast majority of the journal articles from which I have quoted or otherwise cited were accessed by me online. This had many advantages—in addition to being able to read material from journals not available at my university library—and allowed me to make use of a vastly greater and wider array of sources to bring to bear on the topic that I was exploring.

There have been times when I pursued another author's sources and found errors in the documentation. I have always favored a reference as complete as possible, therefore providing a kind of redundancy that would make it easier for the reader to find the cited source even if there were an error in the citation. For this book, virtually all the reference entries were simply transferred from the library catalog, Worldcat, or the journal in which an article was published. I have kept the redundancy in the documentation even though the new technology reduces the probable error rate to very close to zero. Similarly, wherever possible, I simply copied whatever I wished to quote from the online source as it provides a degree of confidence in the accuracy of the quotation not attainable before.

Many journals give the user a choice of a PDF file or a text file. A text file readily allows one to save the article on a disk for future use and to use this file for quotations and reference citing but it does not provide an accurate page reference. A PDF file replicates the journal page and accurate page citing but denies the user the other advantages of a text file. Many journals do not offer the PDF option for downloading. Consequently, in downloading and saving articles for future use, I was unable to provide the page reference for the quotation when I finally used it. For those who go on line to check my sources, which I assume will be the vast majority of the readers, this is no problem since they will simply search the article using a couple of words from the quote. Out of curiosity, I went to Google and searched a number of my quotes using key phrases. I was pleasantly surprised at how quickly the article came up. My apologies to those who try to find the quotation in the hard copy, but since the articles I used tended to be in the best journals, reading the entire article to understand the context from which the quotation was taken is undoubtedly worth the effort.

Introduction

This book is about two contrasting streams of ideas from the last two centuries in Western thought. On the first is the flow of ideas in chemistry and related areas of biology that has created the conditions for modern medicine, modern food production, and the revolution in biotechnology that is now under way. The second stream is the "vitalist" reaction to the rise of modern science. Vitalism has been and remains at the core of the rejection of modern agriculture in the advocacy of the "organic." Building on the quantitative chemistry of Lavoisier in France, organic chemistry and the vitalist reaction to it arose in Germany though vitalism in a rudimentary form has a history going back to Heracleitus and Aristotle.

When some of us first encountered elements of the history of technology and scientific thought in primary or secondary school, they were always framed in terms of the pursuit of truth, the expansion of knowledge, and the ability of humans to cast out darkness and take command of their destiny. This is the interpretation of science and technology against which postmodernists are rebelling. While nobody holds to this vastly oversimplified view, in broad outline, human inquiry has been making inroads against the darkness of not knowing and has greatly expanded the domain of knowing and understanding. To many readers, I will appear to be a prisoner of this perspective, which with all the caveats and qualifications that anyone would make, I would still plead guilty. The advance of scientific inquiry over the past two centuries has not gone unchallenged from the prevailing ideas that were overturned. For these ideas that were being overturned, we make extensive use of the apt phrase of James Webb, "rejected knowledge" (Webb 1976, 10). Though certain ideas such as vitalism were rejected by the mainstream of science as inquiry proceeded over the last two centuries, this rejected knowledge became central to a stream of beliefs that have been challenging scientific beliefs as they became established and are now at the very core of the contemporary criticism of modernity. Understanding their historical development is helpful if not essential in understanding the basis for the commitment to "alternative medicine" and to everything organic and "all natural." I use the concept of vitalism as an integrating element in this stream of rejected knowledge.

Contemporary vitalist proponents of organic agriculture believe that it must be "pure" since it confers some special kind of virtue both on those who produce it and those who consume it. Harmony, purity, and the heroic are all integral to the antimodernist conception of self and society. Organic agriculture forces the true believer to deny that any evil could ever be involved in it no matter how irrelevant or incidental to the agricultural process it may be. It forces people to deny that there is any use of pesticides in organic agriculture even when reporting studies that clearly indicate the use of so-called natural pesticides. It did not happen because it could not. One can give the URL for the National List of Allowed and Prohibited Substances of the National Organic Program, United States Department of Agriculture, and many still will not believe it (http://www.ams.usda.gov/nop/NationalList/FinalRule.html). When it was free to access, I often gave my students the URL for the organic products site for allowed pesticides and then had them compare the toxicity of some of the approved organics with a synthetic chemical pesticide like glyphosate with a list on the toxicity of chemicals of a respected environmental group.

Organic chemistry and other developments in science and technology allowed the world's population to grow from 1.6 billion to over 6 billion in the course of the twentieth century. In the process, the 6 billion were better fed at the end of the century than the 1.6 were at its beginning. A current standard argument against genetically modified food is that there is enough food to feed everyone were it only fairly distributed. This is ironic coming from some of the same people who not too long ago were forecasting the most ghastly doomsday scenarios of mass famine and death from overpopulation. It is doubly ironic because it is an argument also by those who opposed the science and technology of the green revolution that transformed food availability and accommodated a better than doubling of the world's population (Evenson and Gollin 2003).

Many of the leading luminaries of the anti-genetically modified food movement continue to proclaim the green revolution to be a failure while still arguing that there is enough food for everyone. If the green revolution failed, where does this "enough food for everybody" come from? Nobody claims that we are producing enough to feed the projected future population, nor does anyone have any viable proposals as to how we may do so. Critics continue to vilify those who made the green revolution and those who are working to create a new double green revolution. Organic chemistry, genetics, and now molecular biology have been essential to twentieth-century advances in agriculture, such as plant breeding, and provide a framework for what is needed to keep the process moving forward.

Antiglobalization has combined with the anti-genetically modified food mania in an ideological cluster used to raise money and mobilize protesters into the streets.

I begin this book with an exploration of the factors involved in the modern fear of technology, which forms the foundation for a complex of antitechnology beliefs and practices and leads to seeking alternate lifestyles and the lifeways of other peoples. I will attempt to contrast the history of the sciences involved in modern agronomy and food production with the history of dissenting theories and practices that underlie what is called organic agriculture and alternative medicine in a manner resembling the two strands of the double helix. The two strands of my double helix for the two centuries of the growth of scientific understanding and the vitalist reaction to them are the core contribution of this book. True to the idea of the double helix, I tried to weave the two narratives together but could not do it satisfactorily. I decided on a strategy of less than complete interweaving trusting that the reader would see that the science narrative was largely for the purpose of defining the ideas that vitalism was rejecting and revolting against.

When vitalism was banished from science over a century ago, many scientists assumed that this was the end of vitalism except for a few proponents in philosophy where it died out a couple of decades later. I will argue that they were mistaken and that vitalism is at the core of an array of contemporary antiscience and antitechnology movements. In my judgment, one cannot really fully understand the ferocity with which certain beliefs about homeopathic medicine and the organic are held against all evidence to the contrary, without the historical perspective and understanding of the underlying stream of vitalism and its "rejected" status in terms of modern scientific knowledge. In a careful search of the literature, I was unable to find studies of the vitalist history of the current movements or any that contrast them with the science that they are rejecting. Thus the need for this book. Needless to say, I find James Webb's thesis of rejected knowledge to be extremely useful for understanding contemporary movements and have sought to expand it beyond his use of it to describe the rise of the Nazis in Germany.

Central to the differences between science and vitalism in its various manifestations from Woehler's first synthesized organic compound to synthetic fertilizer in modern agriculture are nitrogen or nitrogen compounds and, most of all, nitrogen as a resource supporting life. Consequently, in this volume, I return in a greatly expanded form to history of nitrogen in life and in agriculture. I integrate it with another recurring theme, namely that technology creates resources. I do not know of any idea more calculated to keep people impoverished than the idea that resources are natural, fixed, and finite. This idea has become dogma among critics of development policies and has led to wasted expenditures and continued calls for more waste in the present. The need for chapter 9 ("Feeding Six Billion People") is compelling in countering myths of scarce resources and beliefs about the sufficiency of organic nitrogen to feed the world's population.

In the epilogue, I devote a few paragraphs to earlier issues when new research has further clarified them. I then close with a positive note about modern science, modern technology, and modern life. In the epilogue, I devote a few pages to science, truth, and beauty illustrating this section with contemporary advances in astronomy. Though this book is largely about science and technology providing us with our daily bread, science and technology are about much more than bread alone.

Origins of
The Organic Agriculture Debate

CHAPTER 1

Science, Technology, and the Critics of Modernity

H istorically, nineteenth- and twentieth-century romanticism in Europe and North America have been seen as a revolt against the Industrial Revolution. But it has also been a re-action against science as presumed dangers of knowledge. The hubris of wanting to know forbidden truths, and the thesis that there are things that people are not meant to know have deep roots and were manifested in the legend of Golem and various iterations of Faust, even before Goethe's rendition. The frontiers of science moved ahead in the nineteenth century pushing back the domain of the arcane and mysterious. Romantics refused to cede this territory.

The Human Machine?

Between eighteenth-century Newtonianism and Darwin, there was another revolution in thought that shaped Darwinism and much of the anthropology that was able to distinguish between myth and magic, for example between garden magic and agronomy. Though a modern engineer may use Newtonian mechanics as matter-of-fact knowledge in a technological endeavor, eighteenth-century Newtonian mechanics, when utilized as social philosophy, lacked the truly revolutionary outcome of removing magic or the belief in the "mystic potency" of unseen forces (Hamilton 1999, 90–117). That was achieved by the revolution in thought in chemistry, laying the foundation for modern chemistry, agriculture, nutrition, physiology, and medicine. And chemistry remains at the heart of many contemporary issues, conflicts, and strange dichotomies, as synthesis (as opposed to reductionism) is good while synthetic is suspect, organic refers to something other than a carbon-based compound, and "chemical" has become a code word for manufactured chemicals and the source of evil in modern life.

Jean Mayer in his Lowell Lecture (1989) dates the origin of scientific nutrition with the work of Antoine Laurent Lavoisier (1743–1794). "First came an understanding of the organism as an **engine.** The understanding of the energetic aspects—the caloric aspect of nutrition . . . started in the 1780s, with a very famous set of experiments conducted in 1789 by Lavoisier," which "established clearly that there was a similarity, indeed, an identity between the phenomenon of combustion and the phenomenon of respiration and that respiration was the oxidation of foods by the individual; that what one observed was in fact a machine, an engine, burning food in order to function, to maintain its body temperature, to move, to grow." Lavoisier's work on combustion overturned phlogiston theory and related theories of the alchemists (Asimov 1962, 48–49; see also Fruton 1999, 234–37).

Many of the basic ideas about animals being like machines can be traced back to Descartes who has been demonized by many contemporary postmodernist critics as a father of reductionism with all of its attendant evils. Reductionism had earlier origins in the 1200s with (William of) Ockham's razor or law—do not posit entities beyond necessity. Ockham's law has been an important element in scientific inquiry ever since.

Lavoisier's work was empirical and quantitative and therefore testable and refutable. Lavoisier and his successors were treating the living organism as an internal combustion engine before it was invented. The quantitative treatment of animal metabolism was prior to the thermodynamic treatment of steam engines beginning with Sadi Carnot (1796–1832) in the 1820s. The study of the heat transfer of engines was heuristic, transforming physics, chemistry, and biology with the work in thermodynamics of James Joule (1818–1889), Hermann von Helmholtz (1821–1894), Rudolf Clausius (1822–1888), Ludwig Boltzmann (1844–1906), and Josiah Willard Gibbs (1839–1903).

The formulation of the second law of thermodynamics by Clausius led to the recognition that entropy not only increases in closed or isolated systems but also is increasing in the entire universe. (The first law of thermodynamics is the conservation of matter and energy.) If heat transfer is the basis for a functioning engine—work—then in time as heat transfers from warmer to cooler objects, uniformity will emerge and no more work can be performed. This was often referred to as the "entropy crisis" and as with most crisis in science, it gave rise to speculative thought, inquiry, empirical investigation, and advances in knowledge. Living systems are what has been called "islands of anti-entropy" as they take in energy (increasing the entropy around them) and build up order and differentiation (Wiener 1989). As long as energy continues to flow from the sun to the earth, then life on earth can continue to build complexity.

The Unity and Beauty of Inquiry

In many respects, the various disciplines of science became intertwined, as advances in one gave rise to new understandings in the others. The work in combustion and spectral analysis of it in chemistry allowed astronomers to determine the chemical composition of stars and other heavenly objects. Increasingly, it became understood that the universe was governed by forces and composed of matter and energy comparable to those on earth. Up to the Renaissance and early scientific revolution in Europe, it was believed that the earth and the heavens were of different substances or essences (with the heavens being the quintessence or fifth essence—the purest and most refined essence—and with the four terrestrial essences being earth, air, fire, and water) and governed by different principles, beliefs that were finally and firmly laid to rest in the nineteenth century. Dimitri Ivanovich Mendeleev (1834–1907) with his periodic table showed that one could impose an order and understanding to the elements. Science then and now has not driven all the unknowns and mysteries out of the universe, and maybe it never will, but it has shown that it has the means to continually push out the frontiers of knowledge and expand human understanding and our ability to function in the world and to improve the condition of the human endeavor.

Mysticism, vitalism, and various contemporary antiscience systems are neither necessary nor helpful and are positively harmful in their opposition to the utilization of scientific knowledge for human betterment. Many have tried to play the role of magus or magician from early shaman to alchemist to contemporary holistic healers but it is the scientists who delivered. Who in 1800 would have dreamed that within the century humans would be able to analyze the elements in celestial bodies and even discover an element, helium, in our sun, which was found on earth over a quarter century later. Robert Wilhelm Eberhard Bunsen (1811–1899), inventor of the Bunsen burner, allowed us to identify extraterrestrial elements with his invention of the spectroscope (Asimov 1962, 83–86). Could anyone in 1800 have guessed that humans would someday be able to understand the forces within our sun and the stars that set them ablaze and light up our lives? Now we know that the very elements that are vital to life as we know it could not have been formed in the big bang, which created hydrogen and helium. The elements up to iron were forged in an earlier star and trapped there until it ended in a violent death called a supernova, which also created the heavier elements. We can view ourselves as being the ashes of dead stars or maybe "each of us and all of us are truly and literally a little bit of stardust" (Fowler 1984). From these ashes, our solar system including earth and all life on it were formed.

Science and the Origins of Life

In recent decades there has been increasing speculation that not only the ingredients of life came to us from outer space but the amino acids themselves were created in space and fell like dust upon our not yet living planet, possibly even becoming the first living matter here (Shock 2002; Bernstein et al. 2002; for some interesting speculations and a survey of some of the current theories on the origins of life on earth, see Wills and Bada 2000 and Bada and Lazcano 2002). If life did not come from outer space, some of the "building blocks needed to start life on Earth may have" in what is suggested as possibly "life's sweet beginnings" (Sephton 2001, 857; Sephton and Gilmour 2001). The recognition of the importance of the sugars in our cells, has many scientists adding a need to understand the glycome as well as the genome and proteome. Polyhydroxylated compounds have been found to be "present in, and indigenous to" well-known meteorites "in amounts comparable to amino acids" (Cooper et al. 2001, 879). "Polyhydroxylated compounds (polyols) such as sugars, sugar alcohols and sugar acids are vital to all known life-forms—they are components of nucleic acids (RNA, DNA), cell membranes and also act as energy sources" (Cooper et al. 2001, 879).

Until a recent study of two meteorites, there has been "no conclusive evidence for the existence of polyols in meteorites, leaving a gap in our understanding of the origins of biologically important organic compounds on Earth" (Cooper et al. 2001, 879). Now having found a "variety of polyols are present in, and indigenous to, the Murchison and Murray meteorites in amounts comparable to amino acids," there is the possibility that some of the vital ingredients for life came from outer space (Cooper et al. 2001, 879). Amino acid molecules have chirality (a property of some "crystals, gases, liquids, and solutions") in that they have no plane of symmetry so that when optically activated, "they will rotate plane polarized light to the left or right" making them L-isomers or D-isomers (Wills and Bada 2000, 264, 15–19). All amino acids in life as we know it on earth are L-isomers, while those found in the meteorites or those synthesized in laboratories tend to be a racemic mixture—containing "exactly equal amounts of the asymmetric forms of an optically activated molecule" so that the mixture "does not cause plane-polarized light to rotate in either direction" (Wills and Bada 2000, 264, 86, 18). All sugars are D-isomers (Siegel 2002). Finding a racemic mixture of amino acids in a meteorite is an indication that the amino acid came from outer space and that the meteorite was not "contaminated" after entering the earth's atmosphere. "Analyses of water extracts indicate that extraterrestrial processes including photolysis and formaldehyde chemistry could account for the observed compounds. We conclude from this that polyols were present on the early Earth and therefore at least available for incorporation into the first forms of life" (Cooper et al. 2001, 879).

There is more than a bit of romance to the scientific understanding of our origins or, more correctly, the variety of possible origins. How did meteorites come to have basic organic compounds? There is the "possibility that they were first formed in interstellar space where there are vast and relatively dense clumps of dust and gas called molecular clouds" (Sephton 2001). "Starlight could have irradiated icy mixtures of water, ammonia and carbon monoxide that coated the surfaces of small dust particles. The resulting reactions may have generated simple sugar-related compounds or their precursors" (Sephton 2001).

Within this icy mixture, a "small dense core" could have been transformed into a "rotating disk of dust and gas which preceded the early Solar System. If simple interstellar organic molecules survived the transformation of the nebula into the Sun and discrete planets, they could easily have been caught up in forming asteroids" (Sephton 2001). Maybe we are all bits of recycled stardust transformed by starlight, and somehow "in the early Solar System, the first chemical steps were taken towards sweet life" (Sephton 2001). This possibility is as interesting and exciting as any tribal legend that humans have ever devised.

Mathematical equations can be elegant or even have beauty (Farmelo 2002). It has been suggested that the human brain is the universe's way of knowing itself. This is another way of stating Einstein's famous aphorism, "The most incomprehensible thing about the universe is that it is comprehensible" (Overbye 2002). From the far depths of both space and time to subatomic particles, from the deciphering of the human genome to the understanding of ecological systems, scientific knowledge is as wondrous and magical, as beautiful and sublime as anything that the mystics and postmodernists have to offer, and science is more useful. Those who call it reductionist, logophallocentric, and a variety of other pejoratives have yet to offer anything better or anything that helps those in need. Possibly, the second most incomprehensible thing is how anyone could find the scope of human knowledge anything but exhilarating and awe inspiring.

CHAPTER 2

Science, Integrated Inquiry, and Verification

To be scientific, knowledge has to be testable and capable of being verified or falsified by finding what our theories predicted. Theories involving cause and effect have consequences that can be predicted and verified. Newtonian mechanics has allowed astronomers to find additional planets in our solar system whose existence could be predicted from perturbations of already known planets. Mendeleev's table could be filled in by later chemists. What Thomas Kuhn called "normal science" is frequently finding specific instances of what was predictable from a theoretical framework whether it be finding new planets, fitting elements appropriately into a periodic table, understanding a disease in terms of a parasitic vector, or later, simply a dietary deficiency. Currently, fifty years of molecular biology from the decoding of DNA (deoxyribonucleic acid) to biotechnology to the human genome project has expanded our knowledge of ourselves and led to a stream of advances in pharmaceuticals and other forms of treatments to heal the sick, extending our lives and well-being. Good science is operational and is at the core of most everything that makes us human.

Vitalism and Verification

Lavoisier's work began the process of freeing science from the vitalist belief in an invisible force or *vis viva*. There is nothing in

Lavoisier or in Mayer or in this view by others, against seeing humans in a multiplicity of other dimensions. Biologists may be "materialists" in denying "supernatural or immaterial forces" and accepting those that are "physico-chemical" but neither do they accept "naive mechanistic" explanations or any belief that "animals are 'nothing but' machines" (Mayr 1982, 52). "Vitalism is irrefutable" and therefore incapable of being considered as a scientific hypothesis or theory (Beckner 1967, 254). Science cannot operate on the basis of a "factor" that is "unknown and presumably unknowable" (Mayr 1982, 52). Nor can it operate with theories that cannot be refuted and therefore cannot be tested. In other words, what David Hamilton calls "matter-of-fact knowledge" is central to scientific inquiry (Hamilton 1999, 90–117).

The understanding of the machinelike characteristics of the living organism was essential for the scientific advances that have given us the longer life and good health achieved over the last two centuries. These advances have furthered the other aspects of the human endeavor in keeping us alive so we can cultivate and appreciate the aesthetic dimension of our being.

Origins of Organic Chemistry (and the End of Vitalism?)

In the eighteenth and early nineteenth centuries, the "apparent uniqueness of life" led to the reasonable belief at the time that there was "something mystical about it, some ineffable force that set it off from the nonliving world." The term *vitalism* was coined in the eighteenth century by Georg Ernest Stahl (1610–1734) (Wills and Bada 2000, 11–12). Stahl's animistic vitalism was no more incompatible with the science of his time than were his theories about phlogiston and combustion prior Joseph Priestly's (1733–1804) isolation of oxygen. Priestley's work was followed in 1828 by the first laboratory synthesis of an organic compound—urea—by Friedrich Woehler (1800–1882), a chemist and founder of organic chemistry (Wills and Bada 2000, 12). He demonstrated that chemistry could create organic compounds even without organic molecules. The prevailing vitalist belief argued that organic molecules could only be formed from other organic molecules. Justus Baron von Liebig (1803–1873), a founder of agricultural chemistry, in his essay "Chemistry in Its Application to Agriculture and Physiology" refuted the theory that only organic material (specifically, humus) nourished plants. Following Lavoisier, Liebig recognized that "respiration involves oxidation of substances within the body for the production of heat" and concluded that the "carbon dioxide exhaled by the body was an index of its heat production" (McCollum 1957, 93).

Among Liebig's most important discoveries was the demonstration that minerals could fertilize soil. The nineteenth- and twentieth-century application of this discovery has allowed a human population six times greater than in Liebig's time to be better nourished than ever before. Liebig used quantitative analysis in the study of biological systems and demonstrated that "vital activity" was capable of being fully understood in physicochemical terminology. His 1840 book *Thierchemie* integrated chemistry and physiology. He showed that plants manufactured organic compounds using atmospheric carbon dioxide. Though the atmosphere has an abundance of nitrogenous compounds, plants could only use those found in the soil. In England, Edward Frankland (1825–1899) developed the concept of valency bonds and the system for writing chemical formulas depicting the bonds between atoms in the molecule (McGrayne 2001, 51). In 1845, one of Woehler's students, Adolph Wilhelm Hermann Kolbe (1818–1884), accomplished the first synthesis of an organic compound (acetic acid) from its elements, which to some observers sounded "the death-knell of vitalism in chemistry" (Toby 2000).

Darwinian Revolution

The Darwinian revolution was clearly consistent with the earlier trends in chemistry and undoubtedly influenced by them. Darwinian theories had to overcome the essentialist beliefs about the immutability of species (Mayr 2001, 78, 83). This was comparable to the challenge to chemistry concerning organic compounds and their essential vital characteristics. Differing from the earlier saltation or instantaneous mutation theories, Darwin, like Lyell, found continuities in a uniformitarian, transformationalist mode of variational evolution of populations (Mayr 2001, 78, 80, 85; Gould 1977, 21). In contrast to vitalist doctrines, Darwinian theory is nonteleological (Mayr 2001, 82). Darwin's theoretical framework is uniformitarian, transformational, and nonteleological. It is truly astounding that, after one hundred years of the Darwinian revolution, molecular biology using DNA for analysis has so solidly confirmed the Darwinian classifications of life-forms by morphology that evolved over the first century of Darwinism. Molecular biology has put the final nail in the coffin of Lamarckianism by showing that "no information can be transmitted from proteins of the body to the nucleic acids of the germ cells, in other words, that an inheritance of acquired characteristics does not take place" in what was called the "central dogma of molecular biology" (Mayr 2001, 85; see also Crick 1958, 1970; Commoner 1968; Fleischman 1970; Olby 1994, 432–34; Judson 1996, 332–33; Keller 2000, 53). Nevertheless, Crick stressed the "central and unique role of proteins" (Crick 1978, 139, quoted in de Chadarevian 2002, 195).

Even if today some biologists further modify the central dogma in recognition of the role of proteins in gene regulation and alternate splicing—a complexity in the sequencing in which a gene could produce more than one protein was recognized by Francis Crick himself—there is still a core belief that all the information for coding and splicing originated in the DNA. In any case, the central dogma has been viewed as an informal set of assumptions shared by those doing molecular biology, and for nearly a half century, it and its many modifications have been heuristic and served the cause of research and development extremely well (Tanford and Reynolds 2001, 239). It is a testament to the power of a theory in science that it gives rise to heuristic research, which continually forces modification of it and may someday even overturn it. Any theory that succeeds a solidly based, widely accepted scientific theory, will come into being because of the work that followed from its predecessor theory. Crick's coining the term *central dogma* is unfortunate because there has been nothing dogmatic about it.

Hooke and Leeuwenhoek and the Discovery of Cells

Technology has continually played a critical role in advancing the life sciences. In the seventeenth century, the microscope allowed Robert Hooke (1635–1703) (*Micrographia* in 1665) and Anton van Leeuwenhoek (1632–1723) to discover cubicles or cells in animal and plant bodies (Garfield 2001, 155; Harris 1999, 1–7, 76–77; Harris 2002, 27–35; and Wills and Bada 2000, 6–11). Hooke found that various items that he studied under the microscope were "all perforated and porous, much like a honeycomb" in which he discovered "cells filled with juices, and by degrees sweating them out" (Janick 2002).

Leeuwenhoek "understood the nature of red blood corpuscles and studied human male spermatozoa . . . (and) was responsible for the first representation of bacteria by a drawing in 1683" (Janick 2002). Until 1838, microscopes were characterized by "chromatic distortion" in the form of "colored fringes of light at the edges of the magnified object . . . because the waves comprising the beam of sunlight passing through their lens was refracted to different extents—producing not one focused image, but superimposed images in red, orange, yellow, green, blue, and violet wavelengths" (Gamwell 2003). "The development of the achromatic lens in the 1830s eliminated this distortion by combining layers of glass with different rates of refraction. Overnight, a crystal-clear window opened into the microscopic realm. For the first time, microorganisms were seen in brilliant natural color and immaculate detail. Earlier naturalists had barely been able to make out cell walls" (Gamwell 2003; see also Gamwell 2002, 45–46).

Lorenz Oken (1779–1851) first argued that life consisted of an agglomeration of "independently viable microscopic units that never arose *de novo* from inanimate matter but always formed by division of pre-existing units" (Harris 1995, 2). "*Nullum vivum, ex ovo! Omne vivum e vivo*" was as "categorical a denial of spontaneous generation as you could want" (Harris 1995, 3). Robert Remak (1815–1865) "provided decisive experimental evidence" that animal cells were "produced by the division of pre-existing cells" (Harris 1995, 10; Lagunoff 2002). The immensely influential Rudolf Virchow (1821–1902) facilitated wide acceptance of the role of cell division with the phrase *omnis cellula a cellula* (Harris 1995, 13). August Weismann (1852–1919) wrote "The Continuity of the Germ-Plasm" expressing the idea that "heredity is brought about by the transference from one generation to another, of a substance with a definite chemical and above all, *molecular constitution*" (Portugal and Cohen 1977, 105).

Gerrit J. Mulder (1802–1880) first identified and named protein, a word drawn from Greek meaning " 'standing in front' or 'in the lead' " (Tanford and Reynolds 2001, 15; and Wills and Bada 2000, 13–14). Jons Jakob Berzelius (1779–1848) "defined catalysis and linked it to the mysterious ferment." The ferments were enzymes but vitalists still held that ferments were the result of vital forces and could only result from living matter. The name *enzymes* was given to them by Willy Kuehne (Lagerkvist 1998, 28). Woehler, who first synthesized an organic compound, was a young collaborator of Berzelius (Lagerkvist 1998, 28).

Toward the Double Helix

As Darwinism was beginning and before the *Descent of Man*, Johann Friedrich Miescher (1844–1895) and his successors were laying the foundation for the creation of molecular biology, which would carry Darwinism and biology forward from the mid-twentieth century onward. Over the next eighty to one hundred years, many of those creating the building blocks of molecular biology would, like Miescher, be chemists or physicists. In 1869, while working in the laboratory of Felix Hoppe-Seyler (1825–1895), a leader in the new field of tissue chemistry at Tubingen, Miescher found phosphorus in human cells, which was later identified as nucleic acid, which he named "nuclein" (nucleic + protein) (Lagerkvist 1998, 44–60; Judson 1996, 11; Garfield 2001, 156). "He was able to show that, like proteins, these nucleic acids were made of simple building blocks" (Wills and Bada 2000, 14). "Proteins are made up entirely of amino acids, but nucleic acids are built from three types of small molecules: sugars, phosphoric acid, and basic compounds that were later identified as purines and pyrimidines" (Wills and Bada 2000, 14).

A coresearcher of Hoppe-Seyler, Albrecht Karl Ludwig Martin Leonard Kossel(1853–1927), analyzed the "constituents of the cell nucleus separating the nucleic acid from the protein and identifying adenine, cytosine, guanine, thymine and uracil" (Tanford and Reynolds 2001, 490). Adenine and guanine are purines and cytosine and thymine are pyrimidines in DNA with uracil replacing thymine in RNA. Richard Altmann (1852–1900) in 1889 isolated and named "nucleic acids" (Lagerkvist 1998, 71). The molecular structure of purines was identified and named in 1898 by Emil Fischer (1852–1919); the name *pyrimidines* was given by Adolf Penner (1842–1909) in 1884 (Portugal and Cohen 1977, 69).

Chemistry: From Dyes to Drugs and Other Discoveries

William Henry Perkin's discovery of the aniline dye color mauve, in 1856, "changed the world" as one biographer put it (Garfield 2001). Perkin and others were building on the work on benzene by Michael Faraday in the 1840s. In our history books, we have all heard of the "mauve decades" as Perkin's aniline dye influenced fashion and the arts in general. Only a cynic or an unredeemable technophobe can not marvel at Perkin and his successors being able to take this dirty, filthy substance, coal tar, and create an array of beautiful colors that were far cheaper than the vegetable dyes that often could only be afforded on clothing for the rich. Perkin was trying to synthesize quinine when he accidentally discovered the mauve color. In seeking and failing to synthesize quinine for use in treating malaria, Perkin gave his fellow researchers a magnificent tool for the very scientific inquiry that advanced our knowledge of the human organism, human disease, and would in time even provide us with life-saving pharmaceuticals. Until then, we had burned coal for warmth, to run the engines that drained the mines, to power the factories and then the trains and ships. In mid-nineteenth-century England, coal directly or indirectly, permeated most every aspect of life including lighting the streets at night while fouling the air that people breathed and taking the lives of loved ones in the mine. Now as aniline dye, it could be used as a stain for slides in microscopes that would be as transformational as any of its better known impacts (Portugal and Cohen 1977, 36).

In 1862, Louis Pasteur "revolutionized hygiene and medicine" discovering "that the invisible substructure of nature contained" both "beautiful little organisms" and "disease-causing microbes" or germs "that had cursed mankind for millennia" (Gamwell 2002)—his germ theory of disease. In less than thirty years, scientists such as Pasteur and Robert Koch (1843–1910) isolated the microbes for "leprosy (1873), anthrax (1876), typhoid fever (1880), bacterial pneumonia (1881), tuberculosis (1882), diphtheria (1883), cholera (1884), and tetanus (1889)" (Gamwell 2002, 45; see also Travis 1989; Brock 1988, 290; Riley 2001, 96).

In the 1870s, using slides stained with aniline dye, Robert Koch was able to identify tuberculosis, cholera bacilli, and bovine anthrax bacillus and show how germs spread between animals and cause disease (Garfield 2001, 9, 156–57). In 1882, Walther Flemming stained cells with aniline dye, identifying chromosomes and mitosis (*threads* in Greek) (Lagerkvist 1998, 61–62). The word *chromatin*, which was "derived from the Greek word for color, *chroma*" described the "vivid color within a nucleus after staining" (Garfield 2001, 156; see also Keller 2000, 163 ff). For some time thereafter, many assumed that chromosomes were the primary genetic material (Keller 2000, 163 ff). In 1887, Edouard van Beneden (1845–1910) determined that each cell carried the identical number of chromosomes except the egg and sperm cells, which had half the number of the rest of the cells (Harris 1995, 18–20; Harris 1999, 160–63; see also Fruton 1999, 390–91).

In 1884, Hans Christian Joachim Gram (1853–1938) developed the famed gram stain, which is still in use today, though the phase contrast microscope in 1930 replaced the staining smears of tissues on glass plates for some endeavors. A gram positive of crystal violet meant that these stains had taken hold while gram negative meant it did not. This differential reaction to the stain had extraordinary implications. If a microorganism invading the body reacted differently to a stain, then it might be possible to devise one that would be lethal to the microorganism but not to the human body that harbored it. This insight gave rise to Paul Ehrlich's (1854–1915) idea of a "magic bullet" and the use of coal tar derivatives to create pharmaceuticals. Ehrlich became a pioneer in both "immunology and chemotherapy" (Garfield 2001, 9, 154–55). Ehrlich realized that the stain "often combined with a substance to form a chemical reaction" (Garfield 2001, 156). Ehrlich's insight and study of living cells led to the synthesis of Salvarsan, which became the main treatment for syphilis before the discovery of penicillin (Garfield 2001, 157–59).

Sulfa, the first miracle drug (the sulfanilamide group, SO_2NH_2), was developed by Gerard Johannes Paul Domagk using the aniline red dye called Prontosil. Another aniline red dye may today offer help for Huntington's disease (Sanchez, Mahlke, and Yuan 2003). Sulfa drugs, or sulfonamides, were effective against puerperal fever, which took such a horrendous toll in young mothers' lives. In countries like the United Kingdom and the United States, from the 1930s to the 1970s, sulfa drugs played a critical role in initiating a twenty-fivefold decrease in maternal mortality. So many advances occurred from the 1930s to the 1960s and the present—improved nutrition, prenatal and obstetric care, closer attention to asepsis in hospitals—that it is difficult to isolate the health benefit of a factor such as sulfa drugs on maternal mortality except that it is likely to have been very significant.

Sulfa drugs were effective against a variety of other scourges including pneumonia and leprosy (Garfield 2001, 158). From the 1930s to the present, a series of aniline compounds were created for malaria prophylaxis—mepacrine, nivaquine, proguanil, and mefloquine (trade name Lariam)—most of which I have taken at one time or another (Garfield 2001, 188). Perkin's quest to synthesize quinine for use against malaria was at last realized.

From Amines to Vitalamines to Vitamins

The amino acid asparagine, was isolated from asparagus juice in 1806 by Louis-Nicolas Vauquelin (1763–1829) and Pierre Jean Robiquet (1780–1840) (Fruton 1999, 357). In 1850, the first amino acid, alanine, was synthesized by Adolph Strecher (Wills and Bada 2000, 12). As with many heuristic advances in science and technology, Koch and Pasteur's work on microbial infection led to excess diagnostic reliance on these understandings. Late nineteenth and early twentieth century research led to the recognition that disease could also be caused by nutritional deficiencies. These efforts culminated in the discovery of vitamins by Casimir Funk (called vitalamines until it was discovered that not all vitamins are amino acids) and Frederick Gowland Hopkins who won the 1929 Nobel Prize for his work (Hopkins 1929; Fruton 1999, 280).

Fungus and Bacteria: From Disease to Drugs

Fungal infestation has been devastating to human crops and food supply as well as causing a variety of infirmities and death in humans. In developed countries the use of fungicides and the routine screening of some susceptible foodstuffs (with allowable tolerances of ten to twenty-five parts per billion), such as groundnuts (peanuts), has virtually eliminated fungal infestation such as aflatoxins from the food we eat as causes of human maladies but they remain virulent in poorer countries without the technology and wealth to control them. Fungi, like other living microorganisms without other means of defense, have to defend themselves either by secreting toxins or being toxic to potential invaders such as bacteria. Being toxic is not always an absolute, as what is toxic to one species may be a harmless substance or even a digestible protein to another. As with the case with the aniline dyes, the trick was to find a fungus that was lethal to a bacterial infection but relatively harmless to humans. The antibacterial or antibiotic properties of penicillin (*Penicillium notatum*) were discovered in 1928 by Alexander Fleming. The standard account has Fleming returning to petri dishes where he had cultured the "pathogenic bacterium, *Staphylococcus aureus*," and found that a colony of fungus had inhibited the growth of the bacterium (Moore 2001, 71). In the 1930s, Howard

Walter Florey and Ernest Boris Chain purified penicillin (Moore 2001, 71–72).

For over two centuries prior to the use of *Penicillium notatum*, another fungus (or an extract from it), *Claviceps purpurea* had been used by midwives in Europe to treat postpartum hemorrhage, which has long been and remains a major cause of maternal mortality (De Costa 2002). *Claviceps purpurea* is the fungus that infects rye (and other grains such as wheat, barley, and oats) and has caused the excruciatingly painful malady in Europe that was called Holy Fire or St. Anthony's Fire (De Costa 2002). The fungus secretes an alkaloid toxin, ergot, which infects the flowers of certain cereals and grasses, with the grain produced by each infected flower replaced by a black "ergot." The secretions contain toxic alkaloids, which are found in flour made from ergot-contaminated rye or wheat, which when consumed cause a constriction of the blood vessels. This made it effective later, in a purified form, for treating migraine headaches caused by an expansion of the blood vessels. Holy Fire or St Anthony's Fire, was characterized by "intense burning pain and gangrene of feet, hands, and whole limbs, due to the vasoconstrictive properties of ergot. In severe cases, affected tissues became dry and black, and mummified limbs dropped off without loss of blood. Spontaneous abortion frequently occurred" (De Costa 2002).

If the above was not bad enough, De Costa adds, "Convulsive ergotism was often accompanied by manic episodes and hallucinations, especially a sense that the subject was flying; these symptoms were due to serotonin antagonism by various components of ergot related to lysergic acid diethylamide (LSD). The gangrenous and convulsive forms of ergotism could occur concurrently" (De Costa 2002).

In 1935, the active agent of ergot was extracted, named ergometrine, and "given intravenously or intramuscularly both prophylactically and for treatment of postpartum haemorrhage" (De Costa 2002). A variety of synthetic prostaglandins are now also used. As noted above, there were a number of factors contributing to this extraordinary decline in maternal mortality with ergometrine along with sulfa being among the most important for initiating this decline in maternal deaths (De Costa 2002). Unfortunately, the maternal mortality rates remain high in developing countries. Many die from postpartum hemorrhage whose lives "might be saved by judicious use of a few grains of the extract of ergot of rye" (De Costa 2002).

The life-saving and pain-reducing use of ergot is another example where humans take that which threatens and afflicts us and transform it to that which helps us. As John Dewey states it, science and technology are the means developed by humans to take that which threatens us and transform it by turning the "powers of nature to account" to advance the human endeavor (Dewey 1929, 3). It has become faddish to focus on the death-dealing arsenal that the misuse of science and technology has created to further our inhumanity to our fellow humans. Important as it may be to understand and counter this evil, it does not further human well-being by dwelling on it to the exclusion of the vastly greater amount of good that has resulted from human inquiry. Understanding the good that science and technology can do provides the best guide and pathway to its proper use.

In 1943, streptomycin from soil actinomycete *Streptomyces griseus* was the first of a series of antibiotics from bacterium (Moore 2001, 76). Selman Waksman won a Nobel Prize in 1952 for his discovery of streptomycin. For centuries, tuberculosis, or consumption, was the dread disease that wasted humans, and the prolonged death from it was a tragic theme in literature and opera, particularly in the nineteenth century. With streptomycin, there was for the first time an effective medication for tuberculosis (Riley 2001, 102–103). Most of the world's antibiotics are from this bacterium. Its genome has now been sequenced, giving promise of even more powerful antibiotics and the ability to create new ones to stay ahead of the disease microorganisms developing resistance to them (Bentley et al. 2002; Mayor 2002).

Prior to antibiotics, any minor break in the skin could lead to a life-threatening infection. In December 1940, an otherwise healthy man, Albert Alexander, was in the hospital with an infection from a skin lesion, which by some accounts was caused by being pricked by a rose thorn a month before. As he was near death, he was given penicillin, which had not yet been proved effective. The next day he was sitting up in bed but soon suffered a relapse. Since only one gram of penicillin existed at the time, after being administered, it had to be harvested daily from his urine and recycled, a losing battle that ended in his death on March 15, 1941 (Moore 2001, 69–71). Today thirty billion grams (close to five grams for every human) are produced every year (Moore 2001, 71). Penicillin had proved its worth but to be the miracle drug of World War II, it required mass production by deep vat fermentation (using corn steep as feed stock), which was achieved by chemists working for pharmaceutical companies in the United States (Moore 2001, 72–73). Meanwhile, chemists were checking out molds for different species of *Penicillium*, finding *Penicillium chrysogenum* in 1943 on a cantaloupe that had gone bad.

The advances in scientific research from vitamin deficiency to antibiotics remain part of a larger research process that is continuing to improve the life and survival of humans including those most in need. A report that recognizes that there are still countries in the world where mortality of children under five years old remains above one hundred per thousand live births, also notes that research played a vital role in a 15 percent global reduction in child (under-five) mortality in just the last decade of the twentieth century (Dabis et al. 2002). It does not minimize the significance of the countries with the high child mortality rates to note that at the beginning of the twentieth century, most countries in the world, including the United States, had infant mortality (one year old and under) of one hundred or more per thousand live births and child mortality rates of two hundred or more per thousand live births. We sometimes take for granted the advances in science and technology that changed ethical and moral standards of what is acceptable and what is intolerable. A century ago, child mortality rates of well over one hundred per thousand live births were not intolerable; they were about the best that humans could do given the science and technology of the time.

Human life today is better than ever. It offers every prospect of getting better if we have the will to make it happen and the intelligence to understand the power to improve the human condition that is an inheritance we are obligated to continue.

CHAPTER 3

Reductionism: Sin, Salvation, or Neither?

T he life-saving advances in medicine are central to what is pejoratively called reductionism. Reductionism has many different possible meanings, all of which are the object of scorn by those favoring a more "holistic" view of life. Reducing consciousness to biology and then reducing biology to chemistry and physics would be an extreme form of reductionism. Chemists and physicists analyzing the human organism have provided essential knowledge that allows us to understand how the organism works. Reductionism in this sense would mean focusing on specific ever smaller units within a system, in this case, the human organism. It matters little if scientists studying the chemical constituency of a cell believed that this was sufficient to explain the entire organism or was just another piece in a larger puzzle. The program of "experimental biology in the period 1900 to 1953" was to introduce "experimental techniques and fundamental theories of physics and chemistry" into biology. Olby doubts that their intent was to do "away with all biological entities" but only those that were "experimentally untestable" and "quantitatively inexpressible" (Olby 1994, 426). Francis Crick sought a solid scientific basis "to guide the discovery of new knowledge" and to "help us either spot experimental error or to suggest fruitful theories" (quoted in Olby 1994, 426). Crick never "denied the value of studying organisms at higher levels than the molecular" or argued that reducing a "biological entity to physics is to do away with it" (Olby 1994, 425; Castellani 2002).

Certainly, we all favor holistic understandings built upon testable reductionist findings. However reductionist modern science or medicine may be, it will not succeed unless it can explain complexity and the workings of larger entities be they the human body or ecological systems. What we are criticizing here is the assumption of a holistic understanding without a knowledge of the parts that go into creating the whole as well as holism as an ideology opposing modern science and technology (DeGregori 2003). A critical component of reductionism is the belief that scientific inquiry can proceed satisfactorily, and explain phenomena, and develop operational principles without the need for any "vital principle." The history of science over the last two centuries demonstrates that vitalism is an impediment to understanding without any benefit to humanity.

Modern inquiry (including the social sciences and history) is predicated on a diversity of capabilities and perspectives. There are scientists who are admittedly, and are recognized by others as, extremely capable technicians who function best at the "bench" or devising a statistical test for a theory but who are reluctant to venture intellectually beyond the realm in which they are investigating and have detailed experimental knowledge. There are others who are known for their big ideas but often have to work with the more technically oriented researcher so that their ideas can be framed in a manner that fits the criteria of modern science. We live in a very complex world and however reductionist a theory may appear to be, to gain any kind of acceptance it will have to have demonstrated the ability to generate and explain complexity. Little evidence is required for grand and often simplistic holistic theories though their advocates demand that the world operate in terms of them. Few ideas in modern thought are more reductionist than the oft-repeated assertion that modern science is reductionist.

Medicine, Science, and Reductionism

If not reductionism, then what? A variant interpretation of reductionism in medicine would be the specificity of the diagnosis and treatment. Few critics of reductionism in medicine would wish us to return to the more holistic treatment of the whole person by bloodletting to balance the humors or maybe dosing with arsenic. As medicine becomes ever more specific in its targets, toxic or other adverse side effects are less likely. Pharmaceuticals are being designed to use the body's peptide "zip codes" to seek out the cancerous cells in a process called "molecular targeting" and then be able to interfere with their reproduction but not that of normal cells (Abbott 2002). Among these "smart" pharmaceuticals are the angiogenesis inhibitors, which are yet to be proven but show great promise (Veggeberg 2002 41; Pollack 2002). For one new drug, Gleevec, for chronic myeloid leukemia, "even cynics . . . have been taken aback" by its performance. "For cancer researchers, the drug's remarkable success confirms that they are on the right track: understand which genes go wrong in cancer, design therapeutics to correct these defects, and the disease can be beaten" (Abbott 2002, 470; see also Wade 2003). Other new drugs of the same type are Herceptin for breast cancer and Iressa for "terminally ill lung cancer patients." Iressa "targets signaling pathways vital to cell growth and survival. Specifically, it blocks a docking post on cancer cells that receives a chemical signal that triggers out-of-control growth" (Ackerman 2002).

Critics do not understand that specificity and reductionism can only be achieved because the researchers had an understanding of layers and layers of complexity. In fact, "an organism is a complex assembly of different kinds of cells that perform many different functions. A major goal of biological research is to understand how that complexity is generated" (Pawson 2002, xv; see also Oltvai and Barabasi 2002; Milo et al. 2002; Lee et al. 2002). Theories of evolution have to explain the process of reproduction and how the various traits of the parents are assembled to create a complex human being. Reductionism and a truly scientific understanding of complex systems are intricately related in scientific inquiry. If those who contributed specific pieces of knowledge to the puzzle of life were reductionist in the pejorative use of that term, we would still be indebted to them for providing the details for the bigger picture of how the organism works. Fruton rejects the criticism that scientific inquiry is a search for "some exact absolute truth" rather than the provision of "reliable knowledge," which can be examined and used in ongoing scientific research, which increasingly provides "true" representations of "facts" whose "existence is independent of the human mind" (Fruton 1999, 104).

Physics is often considered the most reductionist of modern sciences with some of its leading figures proudly proclaiming their reductionism (Polkinghorne 2002). As with biology, some of us in our first encounter with physics were learning an incredibly simple (reductionist) model that seemingly explained the physical universe. From

John Dalton (1766–1844) to Ernest Rutherford (1871–1937) and Niels Bohr (1885–1962), natural philosophers and physicists had investigated the atom (indivisible) and eventually theorized a structure similar to the solar system with a large mass in the center composed of protons and neutrons with electrons revolving around it. From these components and their various orbits, all the elements were composed and could be thus organized and understood in Mendeleev's (Dimitri Ivanovich Mendeleev, 1834–1907) table. In chemistry, we learned about valence and how the elements were combined to form all the stuff of the material world. Add in Albert Einstein's (1879–1955) $E = mc^2$ to explain the atom bomb (and of course, much more of which we were unaware), and the high school general science student had an elegant model of the universe and seemingly little else to be learned except a few details. At one level of understanding the model is still valid. When I and my fellow students were learning this simple, seemingly complete all-explanatory reductionist model of the atom and matter, those who had formulated it not only were going beyond it, but had been way beyond it for decades with subsequent investigators finding subatomic particles and new complexities and new questions that continue to the present. There are some in physics and cosmology who speak of a theory of everything and the possibility of finding the final answer to age-old questions. The rest of us are content to wish them luck in their quest but are also convinced their findings whether elegant or complex, emotionally satisfying or not, will nevertheless give rise to ever new and exciting questions for a new generation of inquirers to answer.

One historian of science describes his book as being "largely a *history of epistemology* as implemented by scientists . . . in the sense of a *history of questions* and the conditions under which they came to be posed" (Rasmussen 1997, 20). He adds that "no experimental question is posed without reference to the means expected to answer it" (Rasmussen 1997, 20). John Dewey argued that problem stating is the better part of problem solving. Asking good questions, asking the right questions is at the very heart of human inquiry. However many new questions are raised does not lessen the fact that the realm of knowledge and human understanding has been expanded along with the ability of humans to live, act, and expand the fullness of the human endeavor.

In spite of the many ongoing criticisms of reductionism, reductionist science continues to expand our realm of knowledge, and its understandings are producing advances in human health and well-being.

Although science is not the source of sin attributed to it, neither does it make a unique claim to provide salvation. Modern science seeks only to be understood in its own terms as an effective means of understanding the world and allowing us to use this knowledge to advance the commonweal. We can also add that modern scientific inquiry can liberate in other ways in that it can undermine the pseudoscientific mythology that supports unjust discriminatory practices.

In April 2003, many of us celebrated the fiftieth anniversary of the paper by James Watson and Francis Crick in *Nature* (April 25, 1953) on the double helix structure of DNA. To some, this is the ultimate in "reductionist science" in spite of the stream of advances in medicine that have followed from it. To others, the knowledge that has flowed from molecular biology and DNA research offers new possibilities for understanding ourselves and others. The philosopher Meera Nanda suggests that it would be "interesting" to see the reaction of "untouchables" in India (or Dalits—which literally translates "ground down," or oppressed or downtrodden) to the "knowledge that DNA material . . . has the same composition in all living beings, be it Brahmin or bacterium. Or what would a woman do with the knowledge that it is the chromosome in sperm that determines the sex of the new born?" (Nanda 1991, 38).

Traditional beliefs too often found irreconcilable differences between humans and based discriminatory practices upon these alleged differences. It is the modern reductionist science of molecular biology that has literally penetrated to the basic biology of our being to destroy utterly the destructive mythologies of difference while allowing for those differences that help to define ourselves without degrading others. Over 99.9 percent of the human genome is shared by all human beings and of the less than 0.1 percent that differentiate us, only about 3 to 5 percent of that is between groups with about 95 percent being intragroup variation (Rosenberg et al. 2002). The genome that unites us as humans is vastly greater than that which differentiates us, and the portion of the genome that defines our individual biological differences within our culture is itself vastly greater than the minuscule portion of the genome, 0.05 percent, that defines differences between groups (Rosenberg et al. 2002; King and Motulsky 2002; Wade 2002). Call it reductionism if you must, but in my judgment, this vision of the fundamental unity of humans and life in general, when combined with the advances in human life and health, offers the best hope we humans have for our future and for those who come after us.

To critics who raise fears of unknown dangers in modern life, we offer Mencken's comment in another but similar context. "Their effort to occupy all areas not yet conquered by science—in other words, their bold claim that what no one knows is their special province, that ignorance itself is a superior kind of knowledge, that their most preposterous guess must hold good until it is disproved" (Mencken 1930, 311).

CHAPTER 4

On the Trail of DNA: Genes and Heredity

C ritics of scientific inquiry will always point to gaps in a theory such as evolution. Scientists argue that filling in the gaps is where some of the most exciting scientific investigation is being carried out (Holt 2002, 13). There are benefits to science and society from filling in the gaps. "In the history of science, important advances often come from bridging the gaps. They result from the recognition that two hitherto separate observations can be viewed from a new angle and seen to represent nothing but different facets of one phenomenon" (Jacob 1977). "Thus, terrestrial and celestial mechanics became a single science with Newton's laws. Thermodynamics and mechanics were unified through statistical mechanics, as were optics and electromagnetism through Maxwell's theory of magnetic field, or chemistry and atomic physics through quantum mechanics" (Jacob 1977, 1162).

"Similarly different combinations of the same atoms, obeying the same laws, were shown by biochemists to compose both inanimate and the living worlds" (Jacob 1977, 1162). Or as Jacob also states: "Novelties come from previously unseen associations of old material. To create is to combine" (Jacob 1977, 1163).

Those of us who took a high science course before Watson and Crick thought we were learning about how genes determined inheritance. What seemed like an answer to us, was as we shall see, a burning question or set of questions to scientists who used the term *gene* for decades without knowing what it was. In many ways, what a gene is remains an open question though we have many more answers now, which provides us with better questions and enhanced ability to find new answers. For Darwin, the mechanism for inheritance was "gemmules," bits of each part of our bodies that migrated to the reproductive organs where the traits were passed on to the offspring (Portugal and Cohen 1977, 92). From the independent rediscovery of Gregor Mendel (1822–1884) around 1900 onward by Hugo DeVries (1848–1935) of Holland, Carl Correns (1864–1933) of Germany, and Erich von Tschermak-Seysenegg (1871–1962) of Austria, we had knowledge of the outcome of the process. We had yet to unravel how traits were passed on, however (Keller 2000). Even before the unraveling of the double helix of DNA, the advances in genetics were promoting advances in plant breeding (Janick 2002). Even with our vastly greater knowledge today, there are some who are critical of the use of the term *gene*. Enough is known so the concept has "use value," which is "often taken as the goal (and perhaps even the test) of an explanation" (Keller 2002, 343 ff). An explanation "is expected to provide a recipe for construction; at very least, it should provide us with effective means of intervening. Causes, in turn, are identified by their efficacy as handles" (Keller 2002, 343 ff).

A good theory in science is operational. Barry Commoner, in his long-running opposition to the work of Watson and Crick on the structure of DNA and the understanding of replication that has developed over the half century ("central dogma of molecular biology"), has argued that "the inherited specificity of life is derived from nothing less than life itself" (Commoner 1968, 340). While such an explanation may please the believers in holistic inquiry, little can be done with it except use it in opposition to operational use of scientific explanations. Commoner's tirade against the "central dogma of molecular biology" has become the central dogma of those opposed to biotechnology (Commoner 2002).

In 1928, Frederick Griffith published an article on his work with bacteria in which he presented the first indication that the nucleic acid carried the information for inheritance, or the "transforming factor" as he called it (Judson 1996, 17–18; Fruton 1999, 438; and Lagerkvist 1998, 111–13). Griffith died in an air raid on London in 1941. His work at first perplexed others in the field but eventually Oswald T. Avery and coworkers were able to replicate Griffith's experiment (Judson 1996, 18–19). By 1944, Avery was speaking of the transforming principle and was conducting experiments with Colin MacLeod and Maclyn McCarty, colleagues at the Rockefeller Institute Hospital (Hotchkiss 1979). They discovered that the DNA in bacteria transmitted the genetic information. Some doubted the validity of the experiment, believing that a protein contaminant was actually responsible for the transformation and replication. In the first half of the twentieth century, it seemed to many scientists that the nucleic acid was too simple to carry genetic information and something more complex like protein was necessary for the massive amount of information required for inheritance (Portugal and Cohen 1977, IX; Judson 1996, 16–23; Tanford and Reynolds 2001, 229; and Rasmussen 1997, 105). In 1941, George Beadle and Edward Tatum, working with a common mold, stated the dictum, "one gene, one enzyme" (protein) (Lagerkvist 1998, 104–105; Judson 1996, 189; Tanford and Reynolds 2001, 227–29). It was reductionist reducing "genetics to the more manageable proportions of protein science. Conversely, it brought the tools of genetics, transformation of species and the like, to bear on the question of how protein synthesis is managed" (Tanford and Reynolds 2001, 227).

In 1950, Edwin Chargaff studied the base composition of DNA and found that there was a specific relationship between the base pairs that was conserved. The percentage of guanine always equaled the percentage of cytosine and that the percentage of adenine always equaled the percentage of thymine. This means that the purine adenine always pairs with the pyrimidine thymine and the pyrimidine cytosine always pairs with the purine guanine. What is elegant about this pairing is that with the later development of Watson and Crick's double helix, one strand automatically becomes the template for the other strand.

Though many still held to the more complex proteins being the transforming principle, the race was on during the late 1940s to unravel the structure of DNA. It does not detract from the greatness or elegance of the Watson and Crick discovery to say that, as in all scientific inquiry, their work was built on the many discoveries that preceded them or were being carried out by their contemporaries. Alfred Hersey and Martha Chase in their famous "Waring blender experiment" further demonstrated that DNA was the genetic material by showing that only the DNA of a bacterial virus enters the host and not its protein coat. In the 1940s and 1950s, Barbara McClintock was working on transposable elements, whose full significance was not understood until the 1970s when she won the Nobel Prize for it. The magnificent X-ray–diffraction experiments of Rosalind Franklin and Maurice Wilkins revealed the helix structure. Wilkins shared in the Nobel Prize with Watson and Crick. Franklin died of cancer before she could be considered for one (Maddox 2002, 2003). The electron microscope was vital in demonstrating that nucleic acid rather than protein carried genetic information (Rasmussen 1997, 210, 213–14). "Waves of visible light have an average wavelength of 500 nanometers, whereas electrons can be accelerated to wavelengths on the order of 0.005 nanometer (an average atom is about 0.2 nm wide). Observers . . . with an electron microscope see it magnified 100,000 times greater than Pasteur did . . . and are now able to see viruses . . . invisible to Pasteur" (Gamwell 2003).

The story of James Watson and Francis Crick and their unraveling of the double helix structure of DNA is well known and does not need any detailed repetition here. Their two very short papers in *Nature*, April 25 and May 30, 1953, transformed the debate and inquiry on inheritance. Replication was by duplication of the double stranded nucleic acid. With transcription, a strand of DNA (a gene) is read and transcribed into a single strand of RNA as the RNA then moves from the nucleus into cytoplasm. The modesty of the manner (in contrast to Watson in his later autobiographical writing) in which they presented what were truly pathbreaking ideas has often been commented up and illustrated with quotes such as the following: "We wish to suggest a structure for the salt of deoxyribose nucleic acid (D.N.A.). This structure has novel features which are of considerable biological interest" (Watson and Crick 1953a). "It has not escaped our notice that the specific pairing we have postulated immediately suggests a possible copying mechanism for the genetic material" (Watson and Crick 1953a).

In the second of the two papers, Watson and Crick are more confident in their theory but they are certainly not overly assertive and remain more on the modest side of the ledger. "Previous discussions of self-duplication have usually involved the concept of a template, or mould. Either the template was supposed to copy itself directly or it was to produce a 'negative' which in its turn was to act as a template and produce the original positive again. In no case has it been explained in detail how it would do this in terms of atoms and molecules" (Watson and Crick 1953b).

Watson and Crick then proceed to present their model of deoxyribonucleic acid these difficulties with a "*pair* of templates, each of which is complementary to the other" (Watson and Crick 1953b). In order for the process to get under way, the hydrogen bonds have to be "broken and the two chains unwind and separate" before duplication (Watson and Crick 1953b). "Each chain then acts as a template for the formation on to itself of a new companion chain, so that eventually we shall have *two* pairs of chains, where we had only one before. Moreover the sequence of the pairs of bases will have duplicated exactly" (Watson and Crick 1953b).

The next to last paragraph of the second paper begins by admitting that the "general scheme that they have proposed for the reproduction of deoxyribonucleic acid must be regarded as purely speculative" (Watson and Crick 1953b). The rest of the paragraph indicates the vast array of questions their theory raised, many of which occupied researchers like Crick himself over the next decade. We can't repeat too often that it is important for good scientific inquiry to raise new questions as well as provide answers. Again they close with an understatement: "Despite these uncertainties, we feel that our proposed structure for deoxyribonucleic acid may help to solve one of the fundamental biological problems—the molecular basis of the template needed for genetic replication. The hypothesis we are suggesting is that the template is the pattern of bases formed by one chain of the deoxyribonucleic acid and that the gene contains a complementary pair of such templates" (Watson and Crick 1953b).

As with all great advances in science, the double helix raised as many questions as it answered and opened exciting new lines of research, each of which raised even more questions. A half century later, there is an ongoing stream of new answers and a never-ending stream of new problems to be solved. Each new answer raises new questions, adding to the sum total of human knowledge and understanding that have enriched the human endeavor.

Good operational theories or operating principles used in quality research are inevitably modified in use. From the Watson/Crick publication on the structure of DNA, there were questions as to whether it was possible for the strands to separate without breaking even with a weak hydrogen bond. The question was answered affirmatively in 1957 by Matt Meselson and Frank Stahl in what has been called "the most beautiful experiment in biology" (Holmes 2001). For the central dogma, the "first wrinkle on the face . . . came in 1959, when Francois Jacob and Jacques Monod introduced a distinction between 'structural genes' and 'regulator genes'" (Keller 2000, 55). By the early 1970s, Crick had further modified the original "central dogma," which had information flow only in one direction from DNA to RNA (ribonucleic acid) to proteins in recognition that RNA can reverse the information flow from RNA back to DNA. As Francis Crick stated it in 1957, "DNA makes RNA, RNA makes protein, and proteins make us" (Crick, quoted in Keller 2000, 54). By 1977, Richard Roberts and Philip Sharp discovered that the messenger RNA edits out portions of the gene that do not code for any amino acids. In 1978, Walter Gilbert recognized RNA's role in "alternative splicing," which became important in the genome project to explain the process by which a lower number of genes than expected could express a larger number of proteins. Roberts and Sharp won the Nobel Prize in 1993 for their discovery. The portions of genome sequence that code for amino acids are now called exons and those that do not are called introns.

With the double helix demonstrating a mechanism for replication, there was still the long-standing question as to how a code of four letters could create the complexity of the twenty amino acids known at the time. For most of the previous half century, many researchers favored protein as the bearer of inheritance because of its greater complexity, but many, as we have seen, refused to accept the emerging evidence for the nucleic acid believing that experiments such as those of Avery, MacLeod, and McCarty must have been contaminated with protein.

From DNA/RNA to Biotech

During the 1950s and early 1960s, it was recognized that proteins were not made directly from DNA to RNA and from RNA to protein (leaving DNA intact). This is the central dogma triplet of molecular biology (Judson 1996, 281). The RNA sequence is translated into a sequence of amino acids as the protein is formed. Ribosome reads three bases (a codon) at a time from the RNA and translates them into one amino acid.

With so many exciting questions, the discoveries came quickly. From 1962 when Watson, Crick, and Wilkins won the Nobel Prize, twelve have been awarded for work involving nucleic acids. Included in this number were Marshall Nirenberg, Robert Holley, and Har Gobind Khorana in 1968 who were among those finding the codon of three sets of basic pairs that allowed the four bases to combine in sixty-four different ways to sequence to form the protein of life. Holley was the first to do a complete sequence of a nucleic acid, "the alanine transfer RNA of yeast" (Harris 1995, 155).

On the practical side, there was the discovery in 1971 of restriction endonuclease (enzymes) to cut DNA and insert another gene and create transgenic organisms. Paul Berg used this technique to genetically engineer a molecule, which was followed by Stanley Cohen and Herbert Boyer transferring a single gene using plasmids to create a transgenic organism (a bacteria) (Harris 1995, 156–71). The pace of change was so rapid and its potential so enormous that scientists themselves were concerned that the science and technology not outrun our ability to use it wisely, constructively, and ethically. This inspired Paul Berg, Nobel Prize winner and biotechnology pioneer, to call for a moratorium on research until a conference could be held to explore its implications and set guidelines for its use. The famous Asilomar Conference (technically, the National Academy of Sciences' Conference on Recombinant DNA) in February 1975 in Pacific Grove, California, brought together leading scientists to explore every aspect of the issues involving biotechnology (Feldbaum 2002). Asilomar has been an ongoing discussion that continues down to the present. Research protocols were drawn up and research has continued with a flow of beneficial genetically engineered products beginning with the approval of transgenic insulin in 1982. Discussing the moral, ethical, and societal implications of science and technology is as never-ending a process as is science and technology. The NGOs whose 2002 scare campaign against GM corn made famine relief in Africa more difficult and probably led to needless deaths could have used similar meetings to discuss the implications of their actions as could the Europeans whose cynical protectionism on this issue further complicated relief efforts (Paarlberg 2002).

The two centuries of science and technology since Lavoisier has been a time in which life expectancies have doubled and the developments that we have described above along with a variety of others have played an essential role in this transformation. In the process, vitalism has been driven from scientific inquiry. One of the most common uses of the term *reductionist science*, particularly its use as a pejorative, is the idea that science can allow us to explain the functioning of biological phenomena simply in terms of basic principles of biology, chemistry, and physics without a need for reference to any vital principles.

Science Unmodified by Culture, Religion, or Gender

Morris Dickstein argues that postmodernist, poststructuralist theories took hold as a reaction by those who in the late 1960s and 1970s saw the defeat of their earlier idealistic attempts to change the world just as the postmodernist ideas first arose in France following France's defeat and occupation by the Nazis in World War II (Dickstein 2003, B7). In a more critical tone, Meera Nanda describes the reaction of those who have seen ideas go down to defeat that they once believed to be on the verge of triumphing. "What do left intellectuals do when they know that they are too marginalized to change the world? They get busy interpreting the world, of course. And interpreting how we interpret the world, and how the non-Western 'Other' interpret it, and how we interpret others' interpretations . . . and ad infinitum" (Nanda 1996).

Twenty-first-century science simply has no need for vitalist principles. Two centuries ago, one could speak about Western science or Hindu science or any other qualifying definition of science because the vitalism in a body of science was often tied to the dominant religious beliefs of a culture. Today, science and technology need no modifiers as they have become universal. A scientist can be Buddhist, Hindu, Christian, Moslem, Jew, or pagan and each can have differing personal religious beliefs about the human being and aspects of life that are not entirely explained by science. They can be united by shared understandings about the nature of nonvitalist scientific inquiry. The global community of scientists provides a model for achieving unity in diversity from which the rest of the world community could learn much that is needed for a more peaceful world.

There are many who, under the rubric of multiculturalism, proclaim "local ways of knowing" as superior to modern science and promote it for educating different ethnicities or for women's studies (Olson 1999). As a development economist, I greatly value local knowledge such as what the farmer knows about growing a crop or otherwise believes about agricultural practices. It is essential; my work would be impossible without it. We must distinguish between genuine local operational knowledge and the mythical conceptions offered in the name of an ideology.

From 1948 with the election of the National Party in South Africa to the early 1990s, a similar reverence for "local ways of knowing" appropriate to the culture was proclaimed and promoted as "Bantu education." It was called apartheid and many of us spent most of our adult life in active opposition to it. Among the many reasons for opposition to apartheid and its repressive policies was that the so-called Bantu education would handicap the student even in a non-Apartheid society in not providing her or him with the knowledge necessary to survive economically. Today we have what is misnamed as "Science Studies" promoting "Navajo ways of knowing" (which is "assuredly more spiritual and holistic than European ways") in learning mathematics by "teaching calculus before fractions." Among many problems with this procedure would be the "difficulty of expressing the slope of a line, one of the fundamentals of calculus, in any way other than by using a fraction or decimal" (Olson 1999).

From the elite precincts of Western universities, multiculturalism has spread to other parts of the world. Pakistani proponents of "Islamic science" and "Islamic epistemology" have been "citing the work of feminist science critics in their campaign to purge many Western ideas from the schools, and certain feminist professors in the West—perhaps caught up in the thrill of having their work cited half a world away— have favorably cited the Islamicists right back" (Olson 1999).

In the United States, there are advocates of "feminist algebra," whatever that means (Campbell and Campbell-Wright 1995; Levitt 1999, 349; Gross and Levitt 1994, 113–17). When the right-wing Bharatiya Janata Party (BJP) came to power in Uttar Pradesh, India, in 1992, they sought to awaken "national pride" by making "Vedic mathematics compulsory for high school students" (Nanda 1996). "Hindu ways of knowing" involved "government-approved texts" replacing "standard algebra and calculus with sixteen Sanskrit verses." Leading "Indian mathematicians and historians" examined the verses and found "nothing Vedic about them" finding them a "set of clever formulas for quick computation" and not a "piece of ancient wisdom" (Nanda 1996). To Meera Nanda (1996), "in the name of national pride, students are being deprived of conceptual tools that are crucial in solving real-world mathematical problems they will encounter as scientists and engineers." Hinduization extends beyond mathematics to promoting the "Aryan race" coupled with a disdain for all "foreigners including Muslims." The BJP along with the VHP (Vishva Hindu Parishad or World Hindu Council) are offspring of the RSS (Rashtirya Svyamsevak Sangh or Organization of National Volunteers), which has been active promoting hatred of Muslims and Christians in India and has been involved in the destruction of Muslim and Christian places of worship and fostering deadly riots against non-Hindus. The postmodernist/ecofeminist multiculturalism, itself a worthy endeavor, when integrated with a "suspicion of modern science as a metanarrative of binary dualism, reductionism and consequently domination of nature, women and Third World people" supports "Hindu reactionary modernists" who claim the "same holist, non-logocentric ways of knowing not as a standpoint of the oppressed but for the glory of the Hindu nation itself. . . . Re-enchanted science leads straight to Hinduism-as-science (Nanda 2000, 2002). Nanda refers to the "postmodernist-Gandhian-ecofeminist alliance" and notes that "a leading ecofeminist–Vandana Shiva—has become a light of Hindu ecology and makes regular appearances in neo-Hindu ashrams in North America" as well as meetings in India with "RSS chief, Rajendra Singh." "Her work is most respectfully cited in *The Organiser*, the official journal of RSS" (Nanda 2000; Nanda 2001, 185, f.16).

In chapter 10, I discuss the Chipko movement in the Himalayan Garhwal region of Uttar Pradesh, India, for whom Shiva presumes to speak and for which she has won international acclaim. When the Chipko movement's battle for local control of vital forest resources was taken up by Shiva and other "deep ecologists," the local struggle for resources and development was sacrificed to global environmental concerns by groups that "tacitly support coercive conservation tactics that weaken local claims to resource access for sustaining livelihoods" (Rangan 2000, 239). Those who champion local wisdom too often respect it only so long as it is in line with their ideological agenda.

Ideas that are presumed to liberate end up being instruments of oppression. Their advocates in developed countries seem to live in a virtual Potemkin village blissfully unaware that local knowledge and control privileges traditional elites who tend to be dominating males who find the rhetoric of ecofeminism useful but not its desire for equality on classes, races, and genders. Intellectual elites in some developing countries such as Mexico promote practices that are presumed to be based on use and custom (*usos y costumbres*). Since the traditional culture of Mexico was one of male domination, no matter what its intent may have been, the outcome of the use and custom practice is to perpetuate male dominance. It is comparable to outsiders advocating a return to traditional customs and use of the old segregationists' South in the United States, which was definitely not a citadel of rational modern science.

The modernism that opened society and allowed racial and other minorities to demand equal rights and women to challenge male domination is being denied to those who are most in need of change in poorer countries. "The oppressed Others do not need patronizing affirmations of their ways of knowing, as much as they need ways to *challenge* these ways of knowing" as it was the rationality of science and modernity that were instrumental in the creation of more tolerant multicultural societies (Nanda 1996, 1997, 1998, 2003). We can argue how far we have come to a more just society or how much farther that we have to go but it is evident for countries like the United States, the rights of minorities and women have been greatly expanded over the last decades.

A cornerstone in the foundation of modern mathematics is the zero and positional notation, which is believed to have originated in India and spread to the West via various Islamic cultures. This accounts for the fact that terms like "algebra" are Arabic in origin and our number system is called Arabic numerals though they might as well also be called Hindu-Arabic numerals. There are numerous other bits of irony here. Double-entry booking and modern accounting had their origins with the introduction of Hindu-Arabic numerals, which is often dated to 1202 with Leonardo Fibonacci (circa 1180–1250) of Pisa's *Liber Abaci*, which was used as a mathematics text for centuries thereafter. Leonardo traveled with his merchant father to eastern Mediterranean countries where he learned the mathematics of the Arabic and Eastern lands. When Arabic numerals and positional notation first came to Western Europe, their use was banned by many authorities and merchants had to keep secret account books in Hindu-Arabic numerals and then transcribe them into Roman numerals for public accounting.

Some of the twentieth century's greatest mathematicians or scientists in fields such as physics, which are dependent on mathematics, have been from India or elsewhere in South Asia. One of the great mathematicians of the twentieth century was Srinivasa Aiyangar Ramanujan (1887–1920), who was born in the village of Erode in Tamil Nadu, India, when it was a British colony. There is a scholarly journal dedicated to working on mathematical problems identified by Ramanujan. Largely self-taught in mathematics, he was using a half-century-old calculus book, when he came to the attention of the great mathematician, G. H. (Godfrey Harold) Hardy (1877–1947) by sending Hardy some of his work. Instead of the obsolete text, had a text based on the mathematics of the "Aryan race" been available, one wonders whether Ramanujan could have done the great work that he did. Ramanujan's life inspired another young Indian and later Nobel Prize winner, Subrahmanyan Chandrasekhar (1910–1995), to pursue a life in science, becoming one of the greatest astrophysicists of the twentieth century. If the purveyors of Science Studies have their way, Navajo, Zulu, or Maori Ramanujans and Chandrasekhars would not have the opportunity to make their contribution to humankind and receive the proper recognition for it.

What is being promoted as multiculturalism is really an extreme in parochialism. Science is not mechanical but a creative enterprise with an artist-type insight playing a vital role in the hypothesis formulation and the development of advances in theory. A true multiculturalism would occur when all the barriers to participation in the enterprise of science are removed and this endeavor is open to people of both genders and of every race, ethnicity, religion, and any other conceivable heritage and cultural experience that share in our common humanity. Each could bring a differing perspective with multiplicity of differing insights and potentials. In recent decades, we have had a taste, but only a taste, of some of this potential as an increasing number of scientists are coming from developing areas. The disparities and barriers for aspiring scientists from many areas remain an insurmountable obstacle for all but a few. However exciting the differing perspectives may be and however they may contribute to the advancement of science and knowledge in general, one integrating element binds them all together in this supremely humanistic endeavor: the scientific method. It is the great equalizer and the great unifier.

In many respects, I could close this chapter with much the same summation as that for chapter 3. The reductionist scientific inquiry on the trail of DNA, genes, and heredity has provided humankind with a unifying vision of our common inheritance, and the biotechnology that has followed from it offers humankind greater opportunities for health and well-being. In contrast, the promotion of local ways of knowing by outsiders is too often more a projection of the outsiders' own ideology and ends up supporting a discriminatory power structure. On this latter point, I close this chapter with a cautionary tale of how well-intentioned outside intervention can cause more harm than good.

Using some of the same internet networks as those used in promoting the virtues of ideologically based local knowledge, a petition was launched to save a Nigerian woman from being executed for allegedly committing adultery. Some of us who received the petition also received later emails asking us not to sign it. It turns out that the petition was largely phony with numerous errors of fact. Some of those alleged to be signatories had nothing to do with it. More important, those more fully informed on the issues believed that such a petition would do harm to the women's cause (Sengupta 2003). The lesson here as always is that doing good requires that one has genuine local knowledge as to the likely consequence of one's actions. Good intentions are not sufficient (DeGregori 2002, 15).

CHAPTER 5

Vitalism and Homeopathy

E
ven though vitalism has been driven out of science, there are currents of pseudoscience in which it thrives challenging modern science by any means available. A look back at vitalism may help to understand this continuing phenomenon in the face of such success by modern science.

Friedrich Wilhelm Schelling (1775–1854), a student of Immanuel Kant (1724–1804), believed that nature was possessed of a soul as "even inanimate material showed signs of life, as demonstrated by such phenomenon as electricity and magnetism" (Lagerkvist 1998, 25; also, Richards 2002; Harris 2002, 61; Fruton 1999, 115). Everything in the universe was endowed with "polarity" (Lagerkvist 1998, 25). "Life was seen as oscillating between the positive sun and the negative earth" and disease was considered to be caused by a disturbance of the natural polarity" (Lagerkvist 1998, 25, 26). Vitalists "rejected the idea that organic compounds found in the cell could be synthesized in the test tube of a chemist. The formation of such compounds depended on a vital force in the organism. Consequently, a process such as the fermentation of sugar to alcohol was possible only in living yeast" (Lagerkvist 1998, 26).

Louis Pasteur (1822–1895) was about the last great scientist to defend a largely vitalist philosophy. He was working in an area of science, microorganisms, in the late nineteenth century when there was the last glimmer of seeming evidence for a vitalist position. In opposition to Justus von Liebig, Pasteur "could point to a number of soluble ferments, as they were called" (Lagerkvist 1998, 26). For centuries, the astronomer and the astrologer, the alchemist and the chemist, the magus and the healer were often one and the same. Following Pasteur, the final links between magic and science were finally severed. The larger scope of Pasteur's was a major factor in liberating science from mysticism and onto the path of testable hypotheses. Among the many important contributions by Pasteur was his refutation of the theory of spontaneous generation (Harris 2002, 101–24, 132–34; Fruton 1999, 133–43). The spontaneous generation belief had an ironic distinction in that it "could be seen simultaneously as evidence of vitalism and of stark materialism." Since the late nineteenth century, it has been increasingly associated with vitalism (Harris 2002, 319 ff). In the pre-nineteenth-century world "peopled by gods with magical powers, the creation of living organisms from inanimate matter presented no particular problem," but not in a world of increasing scientific understanding (Harris 2002, 1). "As medicine entered the twentieth century, it came to be generally accepted that the atmosphere was indeed teeming with living microorganisms; that infectious processes, whether generative or degenerative, were not the product of intrinsic vital forces or imbalances in these forces" (Harris 2002, 134). To Harris, "experimental evidence in favor of spontaneous generation was hopelessly defective" (Harris 2002, 134).

Vitalism Persists (But Not in Science)

Hans Driesch (1867–1941) in *The History and Theory of Vitalism* refers to the "dynamic teleology of vitalism" (Driesch and Ogden 1914, 6, 46; also Driesch 1908, vol. 2, 340). There is a "real concept of *harmony* in nature, both organic and inorganic" and "*nature is nature for a certain purpose*" (Driesch 1908, vol. 2, 348). Substances have an Aristotelian entelechy or essential nature that works itself out (Driesch 1908, vol. 2). Unlike the trends in nineteenth- and twentieth-century chemistry and biology, issues of morality were and remain an integral part of vitalism. Vitalism even takes on religious overtones in Henri Bergson's *élan vital* (Webb 1976, 17–18; Wills and Bada 2000, 15, 110). In speaking of the "very important relation between morality and vitalism," to Driesch, the "*assertion of morality implies the assertion of entelechy, just as entelechy implies causality and substance. . . .* Vitalism is the high road to morality: morality would be an absurdity without it" (Driesch 1908, vol. 2, 348).

Driesch's idea of harmony is antithetical to the forces that were shaping his era and ours. Harmony or homeostasis are the "anathema of the technological drive which is exploratory in nature. The exploratory drive is rooted in a dynamic imbalance" (Reichel 1983, 495). Earth is the only planet in our solar system characterized by instability and it is the only one that we know for sure has life. "Every living thing uses corrective feedback to sustain the equilibrium of its own unstable self" (Lienhard 1988–1997b; see also Lienhard 1988–1997c). "We humans live in varying states of unstable equilibrium" (Lienhard 1988–1997a). "When we walk, we achieve a brief unstable balance on one foot. Then we begin a non-equilibrium free-fall and catch ourselves on the other foot" (Lienhard 1988–1997a).

What is important is not an inherent harmony but our ability to integrate and coordinate differing functions in order to maintain balance (Smetacek 2002). Smetacek raises the question whether "the mind's balance, and hence its functioning" might not be "derived from that of the body? . . . The human cerebellum, a central organ of balance and also of fine motor skills, contains five times as many neurons as the cerebrum but has received much less attention" (Smetacek 2002). Did the emergence of erect posture and the added disequilibrium in movement evolve along with the brain for managing it and for performing other functions that define us as humans? "As balance is central to every directed movement, evolving fine motor skills is synonymous with fine-tuning the sense of balance. Human evolution can be characterized as stages in differentiation and refinement of our balancing abilities. Our lineage first learned to balance bodies on feet, then tools in hands, and most recently, instruments and aircraft with eyes" (Smetacek 2002).

Smetacek closes his essay with this observation: "The bipedal apes striding across the savannah with head held high evolved a very different proprioception and world view to their slouching cousins. Our ancestors dared to challenge gravity by standing up to it and we continue to do so, with our bodies and tools, minds and machines, balancing our way onwards and upwards, both literally and figuratively" (Smetacek 2002).

An increasing scientific understanding of the world was simply unacceptable to true romantics. Organic chemistry may have sounded the death knell of vitalism but the romantics refused to hear it. For some, the triumph of chemistry made vitalism more imperative. Four centuries earlier, overturning the geocentric view of the universe inevitably brought the astronomers who espoused it into conflict with those whose beliefs and positions of power were predicated upon the Earth being the center of the universe. The inquiry and discoveries of quantitative and organic chemistry were demystifying biological and agronomic understandings and it was equally inevitable that there would be conflict (fortunately, mostly an intellectual conflict) between the emerging science and the defenders of established thought.

Jan Christiaan Smuts: Holism and Evolution

The concept of "holism" was developed by Jan Christiaan Smuts in his 1926 book *Holism and Evolution.* He wrote of "the purity and holiness" of the "human and natural realm." There was a vital principle in all things—energy was a "permanent life-force in the universe." Life originated through an inherent "and permanent life-force in matter that carried out the transformation from the inorganic to the organic world. The 'universal economy,' or harmony, or 'universal cosmic rule' created by the omnipotent life-force was the ultimate foundation for human evolution and the progress of civil society" (Anker 2001, 43).

Smuts rejected "mechanism" using the phrases of modern ecoromantics—"vital," "bioregions," and "thinking like a mountain" with "the mountain" being "the great ladder of the soul" to the highest religion, "the Religion of the Mountain" (Anker 2001, 44, 51–53, 70). "Evolution is marked by the development of ever more complex and significant wholes, rising from the material bodies of inorganic nature . . . to man and the great ideal and artistic creations of the spiritual world" (Anker 2001, 71).

Smuts was strongly anti-Nazi even though some Nazis "may have been inspired by holism, ecology or other 'green' views" yet his ideas about the Bushman being formed by the desert presage similar Nazi views about Jews. Bushman had "become a mere human fossil, verging to extinction." The "confounding factor" was the "semi-desert environment." The Bushman had his "desert nature bred in him." There was "nothing left for him but to disappear" (Anker 2001, 179, 163). To Anker, Smuts and "the South Africans, whose deep ecology terminology of thinking like a mountain endorsed a harsh ecological trusteeship of people of non-European descent, do not really fit the Arcadian ideal" (327 ff).

Aldo Leopold used "thinking like a mountain" as part of his "land ethic." His "land ethic took as its departure the aims and values of his patrons or the 'people with whom I associate' (the ammunition manufacturers). For a nature out of balance he suggested developing rather harsh management schemes by enforcing a 'kill curve' parallel to disease curves for over-populated species . . . 'nothing is gained by killing less than two-thirds' of an animal population to get an ecological effect" (Aldo Leopold at a 1931 conference "representing the Sporting Arms and Ammunition Manufacturer's Institute," Anker 2001, 285–86).

To many believers, Smuts's holism or Leopold's land ethic are ethical and moral principles that confer virtue upon those who hold them and act in terms of them. It might come as shock to many that "thinking like a mountain" to Leopold meant being aware of the long-term damage to it by deer overpopulation, which had to be culled by killing. There is no Bambi syndrome here. Smuts, who could write eloquently in framing the preamble to the United Nations' Charter, could also use holism as a basis for laws oppressing the black population of South Africa. There is neither inherent virtue nor evil in these beliefs except in the use to which they are put. Today, the idea that these beliefs convey special virtue is used by some of their believers to convey a special privilege to violate the rights of others.

The Demystification of Science

By the 1920s, the demystification of science led to a counterattack calling for the "re-enchantment" of science. Max Weber spoke of the "disenchanting" consequence of modern science that undermined "all transcendent principles, systematically stripping the world of all spiritual mystery, emotional colour, and ethical significance and turning it into a mere causal mechanism" (Burleigh 1996, 1404; Harrington 1996, XV–XVI). Reichel shows the continued belief in the "re-enchantment" of science and modern life (Reichel 1983).

Medicine in the 1840s in Germany was the focus of a vitalism under the rubric of homeopathy. Mainstream medicine, with the help of chemistry, was becoming more scientific. Homeopathic medicine was a set of existing practices and beliefs that became part of the vitalist response to emerging modern science that has proponents today using much the same terminology and ideas. Vitalism, holistic healing, and natural cures using herbs and homeopathic medicine are central to "alternative medicine," which devotees prefer to mechanistic modern medical practice.

Homeopathic medicine was and remains a prime area of vitalist principles. Even though its beginnings predated the main work eliminating vitalism from scientific inquiry, it quickly became part of the reaction to the emerging scientific inquiry of the nineteenth century (see for example Evans and Rodger 2000). "Samuel Hahnemann (1755–1843), a German physician, began formulating homeopathy's basic principles in the late 1700s" (Barrett 2001; see also Proctor 1988, 225). The basic principles of homeopathy are "(1) most diseases are caused by an infectious disorder called the psora (itch); (2) life is a spiritual force (vitalism) which directs the body's healing; (3) remedies can be discerned by noting the symptoms that substances produce in overdose (proving) and applying them to conditions with similar symptoms in highly diluted doses (Law of Similia); (4) remedies become more effective with greater dilution (Law of Infinitesimals) and become more dilute when containers are tapped on the heel of the hand or a leather pad (potentizing)" (Mertens 2001; see also Fienberg 2001; Park 2002; Evans and Rodger 2000, 141).

Robert Park checked over-the-counter homeopathic dilutions in health stores and found the dilution listings varied from 6X to 30X to 200C. "The notation 6X means that the active substance is diluted 1:10 dilution in a water alcohol mixture and succussed" resulting in the active substance being "one part in ten raised to the sixth power (10^6), or one part per million." For 30X, it would be "one part in 10^{30}, or 1 followed by thirty zeroes. For 200C, it means that the "active substance is sequentially diluted 1:100 and succussed 200 times" (Park 2002). "That would leave you with only one active molecule of the active substance to every one hundred to the two hundredth power molecules of solvent, or 1 followed by four hundred zeroes (10^{400}). But the total number of atoms in the entire universe is estimated to be about one googol which is 1 followed by a mere one hundred zeros" (Park 2002).

Given the extreme dilution required, Hahnemann "realized that there is virtually no chance that even one molecule of original substance would remain after extreme dilutions." He argued that the "vigorous shaking or pulverizing with each step of dilution" leaves behind a "spirit-like" essence—"no longer perceptible to the senses"—which cures by reviving the body's vital force. In other words, chemical analysis of a vial of a homeopathic medicine and a placebo of plain water and alcohol, "ordinary chemical principles" would not allow one to ascertain "which vial contains the 'active product' and which one the 'placebo'" (Vandenbroucke and de Craen 2001, 508). Some now are even taking the position that continuing testing of the efficacy of "ultradilutions" is "no longer a research priority" and the focus instead should be on comparing the "effect and cost effectiveness of orthodox and homeopathic treatments" (Feder and Katz 2002, 499). This argument is at least a century and a half old when William Bayes maintained in 1856 that the only "value in testing medicines is, *their effect in curing*" as the "vital test" of the medicine itself was beyond the ability of chemistry of the time (quoted in Weatherall 1996, 182). For the ultradilutions ("dilutions of grandeur" in the apt phrase of Whorton 2002, 49) with no molecules of the medicine expected to be left, no chemistry will ever be able to perform this vital test. Evaluating homeopathy without studying its pharmacopeia is like trying to understand Hamlet without knowing the prince. It reduces homeopathy and its alleged efficacy purely to a spiritual or "vital force" (*Lebenskraft*), which, in fact, is what it is.

Modern proponents assert that even when the last molecule is gone, a "memory" of the substance is retained. This notion is unsubstantiated. If it were true, "every substance encountered by a molecule of water might imprint an 'essence' that could exert powerful (and unpredictable) medicinal effects when ingested by a person" (Barrett 2001). A solution is "succussed" when shaken in each step of the dilution (Evans and Rodger 2000, 69–70).

First do no harm is the famed medical saying derived from Hippocrates. Given the extreme dilution, there is a sense in which the homeopathic pharmacopeia meets this criteria since it does nothing. In the era in which it was born, doctors were engaged in giving emetics and purgatives and bloodletting; one could argue, compared to these practices, homeopathy did no harm. The extreme dilution idea was derived from having given patients toxic substances that produced the same symptoms as the disease to be cured. Hahnemann observed that the less of the toxic substance that he gave, the better the chance his patient had to improve. Obviously if you are dosing a patient with toxins, then clearly, less is better than more and nothing is best of all. Hahnemann went from doing harm to doing progressively less harm and calling it help, which was better than some of his contemporaries. But in an age where we have pharmaceuticals that can save and prolong life, doing nothing as a substitute is harmful. Even in the mid–nineteenth century when homeopathy began to flourish, medical practice had been transformed as "heroic bleeding, blisters, and purges had given way to less drastic interventions and milder druggings" (Weatherall 1996, 177).

As would be expected, many who oppose transgenic crops and food production are also supporters of homeopathic medicine. One institutional example would be ISIS (Institute of Science in Society). When there is the slightest evidence of acceptance of homeopathy, the claim is made that they have become "mainstream" (Burcher 2002). A major journal, *Nature*, in the name of fairness, published an article finding some efficacy for homeopathic dilutions because the reviewer could not find any errors in it (Davenas et al. 1988). Given that everything that we know about science is counter to the claims of homeopathy, the editor, in the true spirit of science, invited other researchers to attempt to replicate the results and published the homeopathic article on condition that the senior author, Jacques Benveniste, allow independent investigators to inspect his data (Maddox 1988a,b; Maddox et al. 1988; Benveniste 1988). When no evidence could be found supporting the Benveniste results, *Nature* published the refutation also while the supporters cried foul and claimed a scientific conspiracy protecting the established beliefs (Hirst et al. 1993; Schiff 1995). Schiff refers to work done on the "electrical transmission of chemical information" as a form of medical research (Schiff 1995, 32). Some scientists were critical of *Nature* and its editor for publishing the article (Dixon 1988).

Even if a randomized test of homeopathy appears to produce positive results, there is reason for scientists to doubt it. "Accepting that infinite dilutions work would subvert more than medicine; it wrecks a whole edifice of chemistry and physics. That price is too high" (Vandenbroucke and de Craen 2001, 511). "Too much knowledge that really works in our day-to-day world is built on existing chemistry and physics. We do not want to discard this because of a few randomized trials" (Vandenbroucke and de Craen 2001, 511).

Though we would not wish to discard modern chemistry and physics based on a possible weakly positive experimental outcome, there is an obligation to consider even weak randomized test results seriously, investigate them carefully, and in the case of homeopathy, the virtually certain outcome would be to refute them.

Making Bread from Air

The continued nineteenth-century advance of chemistry, in which German scientists played a leading role, laid the foundation for the early twentieth-century work of Fritz Haber (1868–1934) and Carl Bosch (1874–1940) in the industrial synthesis of ammonia from atmospheric nitrogen, allowing for the mass production synthetic nitrogenous fertilizer. In considering various candidates for the "most important technical invention of the twentieth century," one author, Vaclav Smil, finds none to be as "fundamentally important as the industrial synthesis of ammonia from its elements." It is the "single most important change affecting the world's population—its expansion from 1.6 billion people in 1900 to today's 6 billion—would not have been possible without the synthesis of ammonia (2001, IX; see chapter 9 of this volume regarding the twentieth-century importance of ammonia). The synthesis of ammonia from atmospheric nitrogen was, as some have put it, "making bread out of air" (McGrayne 2001, 58). This invention brought a vitalist reaction in agriculture comparable to the vitalist-homeopathic response with the same connection to present-day movements.

The Vitalist Reaction to Liebig, Haber, and Bosch

To Lady Balfour, a proponent of organic agriculture and a founder of the Soil Association in England, Liebig's "naive theory" that inorganic material could be used in plant production did result in increased food production but it was nutritionally inferior (Balfour 1948, 50–51; Balfour 1976, 56). It lacked a "vital quality," as the modern world, "largely ruled by chemistry," had neglected the "continuity of the living principle in nature" (Balfour 1976, 25). Like Hamilton's mystic potencies, vital qualities were in the eye of the beholder.

Before modern chemical pesticides were an issue, the foundation of organic agriculture for Rudolf Steiner was opposition to synthetic fertilizers, since they were "man-made" and alien to the environment and most of all because they were "dead" (Bramwell 1989, 20; Steiner 1958; Ferguson 1997). As absurd as homeopathic medicine may be, it is topped by what one may call "homeopathic manuring." Similar to homeopathic medicine, Steiner and his followers wanted manure to be diluted to the minutest level in preparations made from rainwater and cow dung that had been buried in a cow horn over the winter (Kolisko and Kolisko 1946, 220–36; Kolisko 1938). In 1919, Steiner founded the mystic cult of anthroposophy encompassing anthroposophical medicine, biodynamic farming, and a mode of teaching stressing art, drama, and "spiritual development." It is still embodied in the Anthroposophical Society's Waldorf Schools. For Steiner and advocates of biodynamic organic farming, the crux of organic farming was nonuse of man-made fertilizer (Webb 1976, 71). Two assumptions constituted the basis of such nonuse: that artificial fertilizer was alien to the environment and that it was dead.

Contemporary proponents of homeopathy still follow the spiritualism of Steiner and believe in such things as the "etheric" and "astral" body (Webb 1976, 62–72; Evans and Rodger 2000). Homeopathy is now part of a largely vitalist movement under the general rubric of alternative medicine. Sometimes the vitalist principle is expressed in scientific, or more accurately, in pseudoscientific terms. One author, Candace Pert, has polypeptides serving as transmitters of vital properties of the mind to every molecule of the body with the result that the "brain is integrated into the body at a molecular level" in what she calls "molecules of emotion" that endow all of life (Pert 1997, 187). Fitzpatrick likens this idea of the "mobile brain"—the "psychosomatic network through which intelligent information travels from one system to another"—to the nineteenth-century "notion of the mobile uterus, which was believed to travel around the female body producing hysterical symptoms" (Fitzpatrick 2002). Needless to say, Pert also believes that the MMR immunization causes autism (Fitzpatrick 2002).

These beliefs may seem strange but harmless. A recent case occurred in Germany where two homeopathic doctors who opposed the MMR vaccine are being blamed for a measles epidemic involving over seven hundred children. "Thirty children have been taken into hospital and the authorities fear there could be deaths if the infection rate continues to rise. The 30 children to be treated in hospital so far have ear, lung and larynx infections brought on by measles" (Hall 2002). "Classical child diseases permanently strengthen the child's immunity and aid progress in the development of the child" is a claim made in a pamphlet circulated in Coberg, Germany, by homeopathic practitioners (Hall 2002). This claim reminds one of the famous dictum of the nineteenth-century philosopher and progenitor of postmodernism,

Friedrich Wilhelm Nietzsche (1844–1900), "What doesn't kill me, makes me stronger" (Nietzsche 1998, 5). The homeopathic practitioners do not seem to know the difference between the building up of the immune system through time by regular minor contact with pathogens and the dangers of experiencing a childhood disease that can have substantial adverse effects. "Their stronghold is the Waldorf School, which actively encourages people not to have their children vaccinated. Now we have an epidemic. The Waldorf School is a holistic teaching centre based on the methods of the late Dr. Rudolf Steiner and is one of several in Germany that promotes alternative medicine. . . . Anti-MMR letters have also been sent to parents by activists advising them not to vaccinate their children" (Hall 2002).

In the United States, a Waldorf School is among those in Boulder, Colorado, where children are not receiving their pertussis and other immunizations with fatal consequences, both for those children not getting immunized and their younger siblings too young to receive theirs (Allen 2002; McNeil 2002). In the United Kingdom in a twelve-month period, "eight infants of preimmunization age have required extracorporeal support for intractable cardiorespiratory failure due to *Bordetella pertussis* infection." Five of them died "despite extracorporeal membrane oxygenation support, and one survivor has substantial neurological disability" (Pigott et al. 2002). Although the reported cases indicate "household infection by infected siblings," parents with infants under five months might be wise to inquire of their New Age/ alternative medicine friends as to whether or not, they or their children have been immunized before allowing them to come over and see the baby.

The British Vaccination Acts of 1840 and 1853 made immunization compulsory. This brought an immediate antivaccination response and movement, which has continued to the present. In 1898, the law was modified to allow exemptions "based on conscience" (Wolfe and Sharp 2002). Ever since, there has been the dilemma between defending community health and the rights of the infant for protection, on one hand, and parental rights, freedom of conscience, and individual liberty on the other.

Tragically, among some groups the immunization rate has fallen below the percentage (generally about 95 percent) necessary for what is known as "herd immunity." This is a level of protection that keeps a disease from becoming endemic and therefore protects the entire community. When only a tiny fraction of the community fails to get immunized, it would be a case of what economists call a "freeloader" problem. But, when large numbers fail to be immunized, it is a threat to the entire rest of the community.

There seems to be a bimodal distribution of children who are not getting immunized in developed countries, namely the children of the very poorest and the children of those with the highest levels of formal education, at least when defined in terms of academic credentials. For the latter, it would be a classic case of what Veblen called "trained incapacity." It takes a considerable amount of what purports to be learning to deny the obvious benefits of immunization (Veblen 1964, 347). Similarly, a study found that infants and mothers were twice as likely to die in home births and were at greater risk for other health problems that could later be fatal. Those choosing to have children at home were highly educated and more "likely to be married, white, non-smokers," which put them in a category that would otherwise be at lower risk of complications (Pang et al. 2002).

There are about 800 Waldorf Schools with about 150 in North America. These anthroposophical centers advocate biodynamic farming, which to some anthroposophists makes typical organic farming look like strip mining. Publication of some of the organic-farming journals of Germany in the 1930s and 40s, resumed after the war, continue to the present. Because agrivitalism is nonfalsifiable, its believers doubt it can be very difficult. The vitalism of Germany, the Soil Association (UK), and modern day organic agriculture have never been scientific. Indeed, they are contrary to nineteenth- and twentieth-century science.

CHAPTER 6

Disenchantment and the Cost of Rejected Knowledge

Enlightenment, Modern Science, and the Holocaust?

R omantic critics of modern science and technology frequently compare aspects of modern science, such as genetic engineering of crops, to what the Nazis did in the 1930s and 40s (see DeGregori 2001, chapter 7, for more details). Undoubtedly, this is the most serious as well as the most erroneous charge against modernism and all it embodies—science, technology, the Enlightenment—that it was responsible for the rise of the Nazi ideology and the Holocaust. The irony of this is that the vitalist/romantic reaction to science and technology had a very powerful presence in Nazi Germany. As one author states, "the Holocaust was driven by a millenarian, apocalyptic ideology of annihilation that overthrew all the enlightened and pragmatic assumptions of liberal modernity" (Wistrich 2001, 239). The forces that shaped the Nazis were complex. The romantic/reactionary/vitalist element was part of the mix of forces shaping Nazi ideology and share a romantic/vitalist heritage with contemporary green movements.

The connecting of modern science, the Nazis, and the Holocaust emerged at the end of World War II, in Frankfurt, Germany, in such works as *Dialectic of Enlightenment* by two exiles from Nazi

Germany, Max Horkheimer and Theodor Wiesengrund Adorno (Horkheimer and Adorno 2000; Diner 1994, 2000, 97–116). This connection is not rare or fading (Friedlander 2002).

In the United States in the 1960s, there was a rediscovery of "reenchantment," "wholeness," and "consciousness expansion" (Harrington 1996, 209). "Holistically oriented German immigrants in the United States, like Kurt Goldstein, Herbert Marcuse, and Fritz Perls, helped teach a new generation of American youthful discontents to speak an individualistic language of wholeness, human potential, and inner transformation, and that this tutelage would bear new fruit in the 1960s and beyond" (Harrington 1996, 210). Harrington adds: "Some advocates of holistic and vitalistic biology . . . are finding a new sort of political and scientific life . . . in the agendas of the ecologically oriented groups like the Green Party" (Harrington 1996, 210).

Romancing and Deconstructing the Environment Nazi Style

The thoughts of Ernest Lehmann, a botanist and an ardent Nazi, sound like some of the most romantic of contemporary back-to-nature environmentalist sentiments. "We recognize that separating humanity from nature, from the whole of life leads to humankind's destruction and to the death of nations. Only through a reintegration of humanity into the whole of nature can people be made stronger" (Lehmann 1934, quoted in Pence 2002, 113).

What romantic imbued with the philosophy of deep ecology could argue with Lehmann's desire to shift the focus from humankind to "life as a whole"? Nor could they find fault with the following until they reached the final three words of the sentence. "The striving toward connectedness with the totality of life, with nature itself, a nature to which we are born, this is the deepest meaning and the true essence of National Socialist thought (Lehmann 1934, quoted in Pence 2002, 113). It is understandable, then, that Michael Zimmerman, a deep ecology theorist, regards Martin Heidegger as a "major deep ecologist" with his "critique of anthropocentric Humanism," his complaints about the destructiveness of industrial technology, his "meditation on the possibility of an authentic mode of 'dwelling' on the earth," and "his emphasis on the importance of local place and 'homelands'" (Zimmerman 1990, quoted in Pence 2002, 136). Zimmerman wisely does find some "problems" with interpreting Heidegger as a "deep ecologist" but keeping these "significant caveats" in mind, he believes that "Heidegger's writings" have much to offer "regarding the environmental crisis" (Zimmerman 1990, 243–44). As with other critics of modernity, Heidegger is interpreted to believe that "humanity should guard and preserve things, instead of dominating them" (quoted in Pence 2002, 136; Heidegger 1970; Heidegger 1977a,b,c).

From the work of the French scholar, Jean-Pierre Faye, Harrington notes (as does Loren Goldner) that the "key words of the vocabulary of postmodernism (deconstructionism, logocentrism) had their origins in antiscience tracts written by Nazi and protofascist writers like Ernst Krieck and Ludwig Klages" (Harrington 1996, 211–12; see also Goldner 2001; Webb 1976, 282). "Dekonstruktion was first used in a Nazi psychiatry journal edited by the cousin of Hermann Goering" (Goldner 2001).

In a discussion of agricultural gene splicing and the Human Genome Project in the June/July 2000 issue of *Ethical Consumer Magazine*, clinical geneticist Michael Antoniou (2000; see also O'Neill 2001) raised the specter of Nazi programs in eugenics. The Nazi specter continues to be raised by those opposed to research on the human genome (Dorsey 2002; Gordimer 2002; Levine 2002).

Science and the Holocaust: The Charge

In *Technoscientific Angst: Ethics + Responsibility*, Raphael Sassower argues that modernity "with its instrumental rationality, its bureaucratic aura, and its cult of efficiency . . . simply extended its tenets and commitments during the Holocaust. . . . Concentration camps are not to be seen as an aberration in any sense of the term: They are extremely disturbing but absolutely rational manifestations of the concern with the Jewish problem" (Sassower 1997, 5).

Sassower cites Robert N. Proctor to the effect that the Nazis did not "abuse" science but rather "contextualized" it in a "vacuum" to fulfill a "specific political agenda" (Sassower 1997, 12). In *Racial Hygiene: Medicine Under the Nazis*, Proctor states: "The Nazis 'depoliticized' problems of vital human interest by reducing these to scientific or medical problems, conceived in the narrow, reductionist sense of these terms" (Proctor 1988, 293; Proctor 2002, 55, is now critical of the Horkheimer and Adorno view of the Nazis). In *Modernity and the Holocaust*, Zygmunt Bauman states that the Final Solution had arisen from "a genuinely rational concern," that it had been "generated by bureaucracy true to its form and purpose," and that "rules of instrumental rationality" were "singularly incapable of preventing such a phenomenon" (Bauman 1989, 17–18). To Bauman, "modern rational society" paved the way for the Holocaust, and anti-Semitism was not responsible for it (Bauman 1989, 89–90). This was also a thesis of the late German historian Detlev J. K. Peukert (died 1990) in a 1988 University of Pennsylvania conference address aptly titled "The Genesis of the 'Final Solution' from the Spirit of Science" (Peukert 1988, 1993). Peukert was the author of *The Weimar Republic: The Crisis of Classical Modernity* (1992) and, until his death, the director of the Hamburg Research Institute for the History of National Socialism. In *Inside Nazi Germany: Conformity, Opposition, and Racism in Everyday Life* (1989), Peukert discussed "pathologies of modernity" and argued that "in the epoch of 'classical modernity,' instrumental reason and the spirit of science assumed hegemonic roles in the ordering of German society" as the Nazis elevated "into mass destruction" the "destructive tendencies of Industrial class society" as part of the "pathologies and seismic fractures within modernity itself" (Peukert 1987, 15). Aly and Heim interpret "rationality" in terms of economic calculus with the Nazi extermination policy based on a cost-benefit analysis (Diner 2000, 138–59, 182–83).

To those who accept Horkheimer and Adorno's line of thought, German anti-Semitism could not have been responsible for the Holocaust, because such causation would have made the Holocaust a unique event instead of an outgrowth of a crisis in capitalism (or modernity)— a "singularity"—that necessitates scapegoating certain segments of the population (for a critique, see Bankier 1994, 118–29). The Final Solution has been compared to the practice of medicine. And Bauman provided a gardening analogy in which the victims of the Holocaust are weeds. The Holocaust, according to Bauman, was a "by-product of the modern drive to a fully designed, fully controlled world" (Bauman 1989, 93). No author, however, has offered in his or her writings any evidence of a causal relationship between science and the Final Solution.

Holism and the Purity of the Past

Those activists who opposed genetically modified crops by "rooting out genetic pollution" in the fields are unfortunately describing their actions in language that is all too reminiscent of the 1930s, though we would not wish to trivialize the latter by comparing them in any other way. Underlying these similarities is a shared sense of the purity of their thoughts and actions and a sense that they are acting in defense of self and higher moral principles. The comparison of modern transgenics to Nazi eugenics is ridiculous on the face of it (Burleigh 2002). The Nazis in all things, race and species in particular, favored the pure over what to them was the impurity of an alien importation. Putting a gene from one species into another would have been as abhorrent to the Nazis as it is today to Greenpeace.

For the Nazis, this sense of purity and the need to preserve and protect it meant that any group, such as the Jews who did not conform to the prevailing beliefs, warranted being eliminated (Pois 1986, 142; Rhodes 1980, 179–80). "No compromise in defense of mother earth" means others are prejudged guilty, so that any action taken that harms them is simply punishment for their sins and can be further justified as being an act of self-defense. An apocalyptic view that the Earth is threatened with imminent destruction absolves one from having to understand the consequences of one's actions, since the good of saving the planet cleanses any wrong that one may do in saving it. Modern neo-Nazis also worry about an "environmental apocalypse," with some calling for an "eco-dictatorship" in taking an "eco-naturalist" position, which "is one that purports to draw normative socio-political conclusions from the findings of ecological science, the lessons of ecology" (Olsen 1999, 45).

Idealizing or romanticizing the past is not only wrong, it can also be dangerous. Robert Pois shows "the myth of the large morally pristine, pre-industrial family" in Germany in the 1930s was both believed and acted upon. Idealizing or mythologizing about nature or other cultures such as the American Indians, has also a potential for unintended harm. "Deep Ecology, biocentrism, nature worship, and New Age paganism reflect a hostility toward Christianity, rationalism, and liberalism in modern society. Although these radical movements have their roots in left-wing dissent, their increasing tendency toward myth and despair indicate their susceptibility to millenarian and mystical ideas on the far right" (Goodrick-Clarke 1998, 231). "A fetishistic cultural pessimism necessitated the creation of a history-defying totemistic past" (Pois 1986, 142). Those who did not fit into the myth were a threat to it in what Olsen called "the darker side of romanticism" (Olsen 1999, 57). The Nazis "exalted synthesis against analysis, unity and wholeness against disintegration and atomism, and Volk legend against scientific truth. . . . Life . . . had an organic unity . . . the invisible force that makes the whole more than the sum of its parts" (Arluke and Sax 1992, 12).

Vitalism, Homeopathy versus Reductionist Medicine

The Nazis also favored holistic medicine and healing and condemned rational scientific medicine as being Jewish and therefore decadent (Aly, Chroust, and Pross 1994, 9; see also Proctor 1988). "The thrust of the Nazi revolution must be to replace the mechanistic thinking of recent medicine by a new and more organic (*biologische*), holistic view of the world" (Proctor 1988, 223). The "natural methods of healing" of homeopathy were a key element in this revolution (Proctor 1988, 223). "The Nazis provided support for areas that today would be considered alternative, organic, holistic, or otherwise heterodox— areas such as ecology, toxicology, and environmental science . . . linked with broader social movements that were trying to reorient German science and medicine towards more natural or 'volkish' ways of thought and living" (Proctor 1988, 224). The medicine that this revolution opposed was viewed as "reductionist" and labeled "Jewish-mechanist thought" (Proctor 1988, 244). Wagner, who was chief physician of the Reich, said of Jewish doctors: "they are sterilizing the medical art and impregnating generations of young doctors with a mechanistic way of thinking" (Aziz 1976, I, 43).

Himmler, Hess, and others were active proponents of homeopathic medicine (Webb 1976, 312). "Natural medicine was not . . . something invented by the Nazis. New was the government's apparent willingness to revive and regulate some of these traditions . . . the attempt to link natural medicine with the ideals of Social Darwinism, racial hygiene, and Nordic supremacy" (Proctor 1988, 226). Unfortunately, what is called Social Darwinism, often has little if anything to do with Charles Darwin and his theories and more to do with Herbert Spencer (1820–1903). In Nazi Germany, Social Darwinism was more the ideas of Ernst Heinrich Haeckel (1834–1919), who coined the term *ecology*, strongly opposed "the mixing of races," favored "racial hygiene," and "founded the Monist League in 1906 in order to popularize the racist version of Social Darwinism among Germans" (Goodrick-Clarke 1992, 13; Budiansky 1995, 56–57). Ecology was to be holistic and monistic or, as Budiansky states it, "rarely has a scientific discipline been so completely defined from its very birth not by what it had discovered but by how it wished to conceive of the universe" (Budiansky 1995, 56–57; see also Chase 1995, 96–97). One might say that Haeckel was a super vitalist or pantheist, as he found vital principles in everything and not just living matter (Harris 2002, 61; see also Chase 1995, 96–97). To Haeckel, nature with its "deep balance and order of the cosmos" provided "nothing less than a handbook for the total restructuring of human society" (Budiansky 1995, 57). His aphorism "politics is applied biology" was apparently repeated by Hitler and Hess (Chase 1995, 126).

German defenders of this alternative medicine poured scorn on scientific medicine and thought in language, much of which has a very contemporary postmodernist ring to it. Karl Kotschau, in a series of articles in 1933–1935, wrote: "In the last hundred years . . . science has turned from 'systems' to 'analysis,' from the recognition of human subjectivity to a belief in 'objectivity' and in a 'science free of suppositions" (Proctor 1988, 164). To Gerhard Wagner, a resolution of the International Medical Congress in Montreux, Switzerland, in 1935 was deemed to be "purely Jewish." A scientist today would find little in the resolution to quarrel with except possibly by what was meant by apolitical. It reads as follows: "Science is simply a matter of truth, and this can never be national. It can only be international, bound to common humanity; science can therefore only be apolitical" (quoted in Proctor 1988, 165). Alfred Rosenberg railed against "hollow internationalism" (Proctor 1988, 224). An essay by Margalit and Buruma frames the same issue in a modern context. "The 'Jewish' idea that 'science is international' and human reason, regardless of bloodlines, is the best instrument for scientific inquiry and is regarded by enemies of liberal, urban civilization as a form of hubris. Science, like everything else, must be infused with a higher ideal: the German *Volk*, God, Allah, or whatnot" (Margalit and Buruma 2002).

Vegetarianism, Organic Agriculture, and Natural Medicine

Vegetarianism and organic agriculture were an integral part of the Nazi ideology for many of its leading advocates. Hitler was a vegetarian as were other elite Nazis who believed in "organic health" (Proctor

1988, 228; Webb 1976, 299, 312; Sax 2000, 35). Rudolf Hess demanded that his food have "biologically dynamic ingredients" (Bramwell 1989, 20; Bramwell 1984, 10). Steiner's influence on Hess and the green wing of the Nazi party was direct while his broader beliefs in the occult may have influenced Nazi theorists such as Alfred Rosenberg indirectly through Russian cultists (Webb 1980, 186; Webb 1976, 309–12). One pair of writers on the Nazis argued that "vegetarianism became the symbol of the new, pure civilization that was to be Germany's future" (Arluke and Sax 1992, 17).

Richard Walther Darre, Nazi minister of agriculture from 1933 to 1943, promoted chemical-free "agriculture according to the laws of nature" and "farming methods according to the laws of life" (Hermand 1997, 53, first quote; Olsen 1999, 76, second quote). The fetish for "organic agriculture" was part of a larger health preference for the natural and concerns about environmental carcinogens, environmental toxins, artificial colorings, and preservatives, and "stressed a return to organic or 'natural' ingredients in pharmaceuticals, cosmetics, fertilizers, and foods" (Proctor 1988, 241, 237). As the U.S. National Academy of Sciences and others have argued, DDT has saved more lives than any other man-made chemical, yet "Hitler's personal physician Theodor Morell declared the pesticide DDT both useless and dangerous and prevented its distribution until 1943 on the grounds that it posed a threat to health" (Proctor 1988, 238). One contemporary academic author who has received numerous prestigious grants to write an award-winning book (*War and Nature: Fighting Humans and Insects with Chemicals from World War I to Silent Spring*), seems to equate using DDT to kill malaria-bearing mosquitoes with using chemical weapons in warfare. At least one sympathetic author interprets him as making this equation (Russell 2001; Sax 2000, 22, 193). After 233 pages of text (with another 66 pages of references to follow), on the next-to-last page of the epilogue, the author does admit that "wars on insects and human beings differed in several respects" even though they are "linked metaphorically, technologically, and institutionally" (Russell 2001, 234). Nevertheless, the author first challenges traditional thinking by putting forth the thesis that "war and control of nature coevolved . . . more specifically, the control of nature formed one root of total war, and total war helped expand the control of nature to the scale rued by environmentalists" (Russell 2001, 2). Russell so assiduously drives his thesis that one author interpreted him to mean that DDT was developed as a chemical weapon for warfare; it was not (Sax 2000, 22,

193). Though Russell never quite says this, it is not only an understandable misinterpretation; given the structure of his argument, it is a likely interpretation.

At the concentration camp at Dachau, "the SS organized farms for cultivating herbal medicines; the project was organized on such a scale that contemporary accounts dubbed it 'the largest research institute for natural medicines in Europe'" (Proctor 1988, 250; Proctor 2002, 55). Himmler supported "a plan for radical dietetic reform." Himmler and others wanted to persuade "people to stop eating foreign spices by replacing them with spices used and produced in Germany. They aimed, too, to encourage the use of 'natural' medicines provided by plants" (Berben 1968, 87).

Himmler's letter of instruction for Dachau and Esterwegen: "I wish the SS and the police also will be exemplary in the love of nature. Within the course of a few years the property of the SS and the police must become paradises for animals and Nature" (quoted in Wolschke-Bulmahn 1994, 145; see also Chase 1995, 124–25). True to Himmler's instructions, "storks were hosted in concentration camps of Dachau and Esterwegen" (Wolschke-Bulmahn 1994, 145). For the families of the German officers billeted at Dachau, life had its privileges. "As Christmas approached, activity in the camps workshops doubled, and then, under a sparkling tree, in a warm family atmosphere, children would receive toys and parents exchange presents made by wretches without a possession in the world, who lay in misery only a few yards away" (Berben 1968, 36).

Dangers in the Belief in One's Own Purity

It is the sense of purity of a past or an organic past and present and a belief that they are responding to a higher calling that I find to be dangerous, whether it is to preserve the purity of the volk or to save planet Earth from contamination from genetically modified crops. These beliefs validate forceful actions that interfere with the rights of others. Genetically modified foods are said to lack vital properties, or prana. Is this a lacking in ceremonial adequacy or is it science?

Scholars like Robert Proctor are often troubled to find some Nazis had favorable beliefs about herbs and organic agriculture while being critical of the use of pesticides. (It should be noted that the pesticides that the Nazis were concerned about were not the modern synthetic chemical pesticides that have been so thoroughly demonized; they were a relative novelty at the time. Rather, the pesticides of the time were arsenic compounds and other natural—derived from plants or minerals—pesticides, many of which are still in use in organic agriculture.) The question is why should they be unless they also have an affinity for these beliefs? The real issue is not whether we share beliefs with purveyors of evil, but what beliefs we share with them and how we share them. It is precisely that vitalism, organic agriculture, and other New Age notions carry with them a sense of the purity of the beliefs for those who adhere to them that causes concern.

Disenchantment and Rejected Knowledge

Those disenchanted with modern culture romanticize and seek to purify the history of other less-developed peoples by denying that they ever engaged in warfare, for example. Even professional anthropologists explained the violence of other less-developed peoples as the product of a Western colonial degradation of a Rousseauian past. Precolonial warfare was denied to these societies "by definition" (Keeley 2001, 332–34). This intellectual self-delusion was furthered by defining real or true war as that between hierarchical armies organized by states in contrast to "the 'ritualized' battles, raids, ambushes, and seldom mentioned 'massacres' of nonstate and prehistoric peoples" (Keeley 2001, 332). Other devices for "pacifying the past involve interpreting" warfare using terms such as *symbolic, ceremonial,* or *economic* and defining "obvious fortifications" and weapons as merely being symbolic (Keeley 2001, 333).

Given these perceptions, it would surprise most of us to learn that the best estimates are that "about 25 percent of the men and perhaps 5 percent of women die from warfare in nonstate societies and that societies in which warfare is not chronic are rare and exceptional" (LeBlanc and Rice 2001, 5; LeBlanc, 1999, 2003; LeBlanc and Register 2003). It is understandable and morally and ethically warranted that social scientists have sympathy for victims of oppression and mistreatment but not at the expense of the truth. Defending the rights of others should not be held hostage to their having a purified societal past unsullied by war or other forms of organized violence nor does the effort to create a more just, peaceful world depend upon an innate human purity and pacifism that has been corrupted by civilization.

The Nazis, or at least a segment of them, embodied virtually every major aspect and theme of the vitalist reaction against the rise of modern science including the condemnation of reductionist science and blaming it for all the real and imagined ills of modern life. Even their virulent anti-Semitism involved criticizing Jews for their modernity. The Nazi are such an extreme instance of the dangers of a constellation of romantic antiscience beliefs that I would not have raised the issue were it not for the fact that contemporary holders of antiscience ideologies have a half-century history of blaming the Holocaust on modern science. Taken in isolation, many of these themes and ideas—purity, hunger for wholeness, romanticizing nature—seem not only benign but downright noble and are so perceived by their contemporary adherents.

It would be wrong to raise the specter of the Nazis against the romantic ideas but it is necessary to indicate the potential dangers even if they are so many magnitudes smaller than those of this chapter as to be not even remotely comparable. Throughout this book, I try to show the many ways that romantic antiscience views such as opposition to immunization can in small or large ways be harmful to humans (for a current example, see Fitzpatrick 2003).

CHAPTER 7

Rejected Knowledge, Nature, and the Occult

Confusion of Myth and Science

Myth, ritual, and superstition have a long history of claiming instrumental efficacy for their beliefs and practices. The garden magic of the Pacific islander was carried out because it was believed that it was as necessary to the production of the crop as was the intertwined practices of cultivation and planting. And the many rituals preceding the hunt throughout the world have been performed because they were believed to be as necessary to bringing down the game as was the hunter's skill. No one could reasonably confuse these ritual practices with modern science in spite of their claims to an instrumental or causal relation with a desired outcome.

Nowhere is the confusion and intertwining between science and myth more apparent than on an issue of relevance both to Nazi Germany and to the present. In most societies, beliefs about pollution, purity, and hygiene have both ritual and operational uses. One can have ritually pure water and food that is in fact contaminated with harmful microorganisms. For many organic food enthusiasts today, toxins in foods or radiation that are deemed to be "natural" are more acceptable than those that are the product of modern industry. The fact that such distinctions may be proffered using the language of science does not make them good science or even a product of science, if they have no basis in fact. Some ideas, such as those of eugenics, may have had advocates who qualified as scientists but clearly, by the time the Nazis were putting them into practice, they were recognized by the scientific community as being bad science.

Nazi Opposition to Humans Dominating Nature

The Nazi ideology had a sense of being in tune with nature, not dominant over it, as is the postmodernist charge against modern science. Hitler clearly stated: "Man should never fall into the misconception that he has risen to be lord and master of Nature . . . rather he must understand the fundamental necessity of the rule of nature and comprehend how even his own existence is subordinated to these laws of eternal struggle" (Dominick 1992, 90). Hitler believed that laws of nature should "guide us on the path of progress" since "the great defect of modern Western society was that man was in constant violation of nature" (Gasman 1971, 162). Hitler in *Mein Kampf* states: "When a man attempts to rebel against the iron logic of nature, he comes into struggle with the principles to which he himself owes his existence as a man. And this attack against nature must lead to his own doom" (quoted in Olsen 1999, 72–73; Fest 1974, 89–106). "He who would live must fight. He who does not wish to fight in this world, where permanent struggle is the law of life, has not the right to exist" (Hitler 1939, 163).

Hitler scorns the idea that "man can control even nature." He says, "there are millions who repeat by rote that piece of Jewish babble and end up by imagining that somehow they themselves are the conquerors of Nature. And yet their only weapon is just a mere idea, and a very preposterous idea into the bargain" (Hitler 1939, 162).

Hitler adds that the "real truth is that, not only has man failed to overcome Nature in any sphere whatsoever but that at best he has merely succeeded in getting hold of and lifting a tiny corner of the enormous veil which she has spread over her eternal mysteries and secret" (Hitler 1939, 162). A man does not create anything but he can discover things. "He does not master Nature but has only come to be the master of those living beings who have not gained the knowledge he has arrived at by penetrating into some of Nature's laws and mysteries" (Hitler 1939, 162). Like a good vitalist, Hitler states that "life-urge of Nature manifests itself" making us "subject to a fundamental law—one may call it an iron law of Nature" (Hitler 1939, 161). "Walking about in the garden of Nature, most men have the self-conceit to think that they know everything; yet almost all are blind to one of the outstanding principles that Nature employs in her work. This principle may be called the inner isolation which characterizes each and every living species on this earth" (Hitler 1939, 161).

Blaming Science and Technology

In addition to blaming science and technology for our modern environmental ills, antimodernists attribute an alleged implicit domination view of science to the Judeo-Christian belief that we were ordained by God to go forth and dominate the world and all who live in it. This belief almost perfectly echoes a thesis of Hans Schwenkel criticizing Jews. "According to the first book of Moses, the Jew does not know nature preservation, because God gave all plants and animals, all that creep and fly, as food to the children of Israel. Only the civilized man, and almost exclusively the Nordic man, gains a totally new relationship towards nature, namely one of reverence, on which nature preservation is also based" (Wolschke-Bulmahn 1994, 143; Fest 1974, 96). "According to National Socialist ideology, an anthropocentric view of nature—that man stands above nature, rather than being simply one, non-privileged part of nature—was to be decisively rejected." Himmler argued that "man is nothing special, only a piece of nature." The predominant idea of "rootedness expressed this idea perfectly, for what was rooted could not be separated from nature" (Olsen 1999, 73).

To attack instrumental rationality as the cause of the Final Solution, some authors use a process of reasoning to attack reason, arguing that the Holocaust was the logical outcome of the logic of modern science. Proctor and others dismiss the argument that what the Nazis did was simply "bad science." I would go further and say the experiments for which they were notorious were not "bad science," they were not science at all. It would be more accurate to say that those who see the Nazis as an outcome of modern science simply do not have an understanding of modern science and technology. If they believe that "chemicals" (code for industrially produced chemicals), modern medicine, and agronomy are killing us, and that this is the hallmark of modern science, then the Nazis do in fact qualify (Proctor 1988, 1999). But if the Nazis were really ahead of their time on vital issues as is alleged, how do they differ on these issues from those who seem to hold similar views? It seems disingenuous to blame the Nazis on modern science and then use as proof of their being engaged in "good science" the fact that they shared beliefs with contemporary critics of mainstream science. I find it absurd if not verging on the obscene to equate the healing powers of modern medicine with the murderous "racial hygiene" of the Nazis.

Nazi Ideologies and Postmodernist Beliefs: The Lineage

The similarities between Nazi ideologies and postmodernist beliefs are the result of a direct intellectual lineage. Keith Windschuttle clearly delineates the Nazi sympathies of the German philosopher Martin Heidegger, which involved his actual membership in the Nazi Party in 1933 to the bitter end in 1945, and continued after the end of World War II, remaining a part of his beliefs until his death in 1976. "Heidegger had sufficient courage of his convictions to republish unchanged, in 1952, his 1935 essay, Introduction to Metaphysics, which refers to the 'internal greatness' of the National Socialist movement, which he saw as a first attempt to come to terms with human fate in the era of 'planetary technique'. . . . Indeed, in his famous 1966 interview with *Der Spiegel*, published upon his death ten years later, Heidegger once again praised Nazism as the first attempt to rethink the human relationship to technology" (Goldner 2001). To an unrepentant Heidegger, the "founding principles of National Socialism" contained an "inner truth and greatness" that provided an answer to the "dreary technological frenzy" that left the world and modern people "destitute" (quoted in Windschuttle 1997, 178, 181; see also Ferry and Renaut 1990, 66; Lilla 2001, 30).

To Heidegger and other Nazi defenders, such as Arnold Gehlen, modernity robbed people of the heroic sense of life, which they and Goering had found in the "heroic German spirit" (Windschuttle 1997, 181). Margalit and Buruma argue that an almost compulsive need to be heroic is characteristic of antimodernist groups of all types and is deeply rooted in most cultures. The ideal is often a "society invigorated by constant heroic violence." The society without heroes is to be despised (Margalit and Buruma 2002). "The common enemy of revolutionary heroes is the settled bourgeois, the city dweller, the petty clerk, the plump stockbroker, going about his business, the kind of person, in short, who might have been working in an office in the World Trade Center" (Margalit and Buruma 2002).

They add: "The hero courts death. The bourgeois is addicted to personal safety. The hero counts death tolls, the bourgeois counts money" (Margalit and Buruma 2002). A. Moeller v.d. Bruck, an early Nazi theorist, wrote that liberalism was "liberty for everybody to be a mediocre man" (Margalit and Buruma 2002). Margalit and Buruma refer to the "sirens of the death cult" for whom the only pathway out of mediocrity "is to submerge one's petty ego into a mass movement, whose awesome energies will be unleashed to create greatness in the name of the Führer, the Emperor, God, or Allah. The Leader personifies all one's yearnings for grandeur" (Margalit and Buruma 2002).

Heidegger was a critic of the globalization of technology to which only National Socialism provided a satisfactory response (Ferry and Renaut 1990, 66–80). Even where Heidegger was critical of the Nazis after the war, he does it by making a comparison that trivializes the Holocaust. "Agriculture today is a motorized food industry, in essence the same as the manufacture of corpses in gas chambers and extermination camps" (Wolin 1993, 290–91; Wolin 1990, 168; Wolin 2001, 3; Ferry and Renaut 1990, 88).

Heidegger was the major influence upon the trilogy of French intellectuals, Michel Foucault, Jacques Derrida, and Jacques Lacan, who were the leading lights of poststructuralism/postmodernism. It takes French postmodernist intellectuals to rationalize the philosophy of someone who ended his lectures with Heil Hitler, wore a Nazi lapel pin, and kept his party membership until the very end of the war (Lilla 2001, 22–23). "No less despicable" was Heidegger's severing "relations with all his Jewish colleagues, including his mentor, Edmund Husserl" and his denouncing, in "secret letters to Nazi officials, a colleague, the future Nobel chemist Hermann Staudinger and a former student, Eduard Baumgarten" (Lilla 2001, 22). There were other connections such as Paul de Man, a leading proponent of poststructuralism/postmodernism in the United States. As a young man in Nazi-occupied Belgium during War II, he had been "a Nazi collaborator and had written a series of articles in Belgian newspapers supporting the Nazi cause" (Windschuttle 1997, 178–80).

Landscape and the Nazi Ideology

If one is looking for an organized, coherent set of beliefs that could be used by the Nazis as a justification for the Final Solution, then the Nazi ideas about landscape and the environment are far better candidates than science and modernity. To the Nazis, "cultures were rooted in the soil, organic as plants. Those cultures, such as those of the Gypsies or the Jews (in short, all 'cosmopolitans') that had no particular landscape or *Heimat* could not really be said to be cultures at all, were rootless, and thus contemptible" (Olsen 1999, 62–63). Hitler considered the Jew to be the "only human capable of adapting himself to any climate" (Gasman 1971, 165). Earlier, Sombart had linked the "wandering Jews" to the desert and modern capitalism while Nordics were people of the forest "attuned to the mysterious, immediate, dreamlike, and concrete. The Nordic community of peasants and craftsmen were 'living, organic . . . and matured.' The brilliant sunlight and clear, moonlight nights of the desert encouraged abstraction and rationality and discouraged 'sense perception and an emotional relation' to inner and outer nature" (Herf 1984, 140). And it was this nomadism of the Jews that "first elevated quantity over quality in economic life. Capitalism had been the product of the 'endless desert' rather than the rooted forest" (Herf 1984, 140). "Rootless" Jews and Gypsies did not belong, but the Germans were "a biologically pure and inviolate race, as 'natural' to its terrain as indigenous species of trees and flowers" (Schama 1995, 118).

In contrast to the Jews, there was believed to be an organic unity between landscape and German history (Wolschke-Bulmahn 1997b, 191). Nations were derived from nature, Germans being a product of a "race-specific landscape" with a "connection to the soil" (Wolschke-Bulmahn 1997b, 190; Wolschke-Bulmahn 1992, 41; Hermand 1997, 53). There was an aesthetic dimension as well, as the ancient German tribes had a "genetically inherent 'feeling' for landscape beauty" (Wolschke-Bulmahn 1997b, 199). Jews and Gypsies were alien to this environment, so they were candidates for extermination. To Reinhold Tuexen, head of the German Reich Central Office for Vegetation Mapping, anything alien had to be eliminated, including alien plants, as it was necessary "to cleanse the German landscape of unharmonious foreign substance" (Groening and Wolschke-Bulmahn, 1992, 123). "In 1942, a team of Saxon botanists, supported by Tuexen, equated the fight against foreign plants with the fight . . . 'against the plague of

Bolshevism'" (Groening and Wolschke-Bulmahn, 1992, 123). It is a testament to the depth of Nazi fanaticism and commitment to a romantic ideology that during some of the worst periods of the war, when Germany was desperately short of manpower, able-bodied men were sent into the forests to root out alien plants by carrying on a "war of extermination" against a small woodland flower, *Impatiens parviflora* (Pollan 1994, 52).

Concerns about alien species in an environment continue to the present and have some legitimacy in terms of potential damage to the ecosystem. Alien species, in addition to our food crops, can also be beneficial, and attempting to remove them can cause harm. In any case, defining a species as being alien can be extremely arbitrary and is often defined as being after 1500 with the European exploration of the globe and the species whose movement they facilitated.

The idea of a landscape specific to certain groups was even applied to Lebensraum. There was a reshaping of the landscape of annexed areas. This has been described as an act of "ecological religion" and "ecohumanism" from which there are "dangers of misuse" when a movement "flees from political reality" (Groening and Wolschke-Bulmahn 1987, 145). These ideas justified displacing others from their environment. "Environmental planners within the Third Reich set about reshaping the Polish landscape" as it was argued that "all Slavs were inherently destructive of their environment" (Olsen 1999, 79). Given their constellation of beliefs, it is understandable that, like the contemporary postmodernists, the Nazis opposed "anthropocentrism." The SS training manual taught that "the concept of humanity is biological nonsense" (quoted in Olsen 1999, 73). The Nazis were critical of the "anthropocentric view of life" (Wolschke-Bulmahn 1994, 143).

The opposition of Nazis to internationalism and universalism resembled the opposition of those who hit the streets to disrupt WTO and other international meetings. Alwin Seifert in 1930 stated that "there exists today a struggle between two opposing worldviews: on one side, the striving for supra-nationality, for equalization of huge areas; on the other, the elaboration of peculiarities of small living spaces, the emphasis which is rooted in the soil" (quoted in Olsen 1999, 53). Seifert favored "biologically dynamic agriculture" (Hermand 1997, 54). Earlier, Ernst Haeckel had argued that "environmental despoliation is the result of the triumph of a universalistic civilization over rooted culture. . . . The conservation of pure nature is seen as dependent on the conservation of the pure—that is, ethnically and/or culturally homogeneous—nation, one that is in accordance with the laws of nature. . . . Defending nature meant defending the homogeneous nation from pollution" (Olsen 1999, 55).

The Nazis, Animal Rights, and Racial Hygiene

There was also strong support for animal rights (Arluke and Sax 1992; Foner 1993). The Führer was touted as the "strongest opponent of any form of animal torment; especially vivisection, which is the 'scientific' torture of animals, and a heinous product of the Jewish-materialistic school" (quoted in Horton 1988, 41). Goering, as administrator of Prussia, issued an edict stating that "vivisection of animals of whatever species is prohibited in all parts of Prussian territory . . . persons who engage in vivisection of animals of any kind will be deported to a concentration camp" (Proctor 1988, 227; Goering 1939, 70–75; Sax 2000, 111–13). The Nazis proudly proclaimed in a party press release in 1933 that "among all civilized states, Germany is the first country to end the shame of vivisection. The New Germany not only frees people from the curse of materialism, egotism and cultural bolshevism, but also gives rights to the tortured, tormented and until now, completely unprotected animals . . . what Reich Chancellor Adolf Hitler and Prime Minister Goering did and will do for the protection of animals, stands as a guideline to the leaders of all civilized states" (Horton 1989, 740; Horton 1988, 41). Attempts to elevate animals often ends up degrading people.

The idea of "racial hygiene" and thinking with one's blood were the foundation of the Nazi ideology (Gay 1968, 82). These ideas were carried over into agriculture and plant life. The German nativists and later, the Nazis, also sought to give the German people a "blood-and-soil-rooted" garden, free of alien species (Pollan 1994, 54; Biehl and Staudenmaier 2000). The "blood-and-soil" doctrine was closely linked by Nazi scientists to ecological concepts that had everything "related to everything else" (Deichmann 2000, 739). Deichmann finds a clear similarity between the Nazi view of ecology promoting the interest of the "volk" and modern environmentalists' "political ecology," which purports to serve the interests of humankind and is critical of nonecological science as being "reductionist" (Deichmann 2000, 739). In his *Conversations with Hitler*, Hermann Rauschning purports to quote Hitler arguing the end of the "Age of Reason" and the beginning of one of "magical explanation." Hitler argued against the idea of there being truth in any "scientific sense" (Rauschning 1940, quoted in Slakey 1993, 50).

Margalit and Buruma ask: "What did Hitler mean by 'Jewish science'?" Or "what explains the deep loathing of Darwin among Christian fundamentalists?" (Margalit and Buruma 2002). "Nazi propagandists argued that scientific truth could not be established by such 'Jewish' methods as empirical inquiry or subjecting hypotheses to the experimental test; natural science had to be 'spiritual,' rooted in the natural spirit of the *volk*. Jews, it was proposed, approached the natural world through reason, but true Germans reached a higher understanding through creative instinct and a love of nature" (Margalit and Buruma 2002).

The Nazis, Vitalism, and the Occult

In the complex set of beliefs in the Third Reich, a larger cadre of key leaders tended to favor extremist environmentalism, organic farming, homeopathy, animal rights, vitalism, occultism, and vegetarianism and to disfavor modern science than has been the case of any comparable regime before or after even in recent cases where the greens have shared power. I challenge the reader to find a contrary instance. These ideas are more appropriate to those who label themselves green as well as advocates of postmodernism, ecofeminism, and/or deconstructionism, than it is to agronomists and genetic engineers. As I repeatedly argue, such proponents are by no means Nazis. Many antimodern critics use the label without any qualifying comments but are indignant if it is used against them even with major qualifications. We can be critical of both their judgment and the validity of the ideas that they espouse. Keith Windschuttle writes scathingly of those "English-speaking academics" who thought they were "participating in an exciting and new theoretical movement" but instead "all they are producing, albeit unwittingly, is an English language version of a French theory from the 1980s, which itself derives from a German thesis from the 1940s and 1950s that was originally developed by a group of ex-Nazis to lament the defeat of the Third Reich" (Windschuttle 1997, 183).

The intellectual thread that links the late nineteenth- and twentieth-century vitalists, the Nazis, the multiplicity of occult belief systems of the last century and a half, and contemporary adherence to vitalist beliefs in medicine and agriculture is what Webb calls "rejected knowledge" and the continuity of specific beliefs he calls the "underground of rejected knowledge" (Webb 1976, 10). In other words, those who find themselves alienated from their society or otherwise reject it would likely identify established knowledge with the established order and therefore turn to rejected knowledge as a basis of their rejection. "It was natural, too, for those elements of society that found themselves excluded by the Established social order to make alliances with the protagonists of an anti-establishment method of thought" (Webb 1976, 10). Germany presented the "unique spectacle of the partial transformation of the underground of rejected knowledge into an Establishment" (Webb 1976, 275). Though changed in the process of becoming established with all the "compromises" being in power demands, "the seekers of 'other realities' did succeed in establishing a different vision of the world from that of the pragmatic, materialist establishment" (Webb 1976, 275–76).

It is extraordinary how many occult beliefs were widely held and adhered to during the Nazi era in Germany. There is a substantial literature on the occultic beliefs in Nazi Germany and their role in the rise of the Nazis to power, so there is no need to revisit them here in any detail. It is not always clear, in reading this scholarly literature, exactly what Nazis believed and how they let certain beliefs influence their behavior. But one repeatedly finds, in addition to Hitler, Himmler, Hess, Darre, and Rosenberg, a number of them (Rosenberg in particular in his *Myths of the 20th Century*) believed in the lost continents of Atlantis and Lemuria, which they got from Madame Helena Petrovna Blavatsky (theosophy) and Rudolf Steiner (anthroposophy). The Atlantis legend included a belief in a succession of races of mankind, the last of which, the Aryans, arose in the cold north of Atlantis and ended up on mountaintops in the Himalayas of Tibet from which they spread to other areas (Webb 1976, 315–16, 499; Spielvogel and Redles 1986). The belief in a Tibetan origin for the pure Aryan Germans was widespread among the Nazis, even among those who did not subscribe to the Atlantis myth (Goodrick-Clarke 2002, 122–24). This Nazi fascination for Tibet has given Hollywood filmmakers some good material for movies. Non-Aryan races were seen as mongrel crosses of the Aryans and local populations. Rudolf Steiner and Madame Blavatsky had followers in other countries so it was not purely a Nazi delusion, but one searches in vain to find any other place where these ideas, flowing from rejected knowledge, had so many followers in power promoting them. However, Steiner has many followers today in the Waldorf Schools and beyond. Many who don't know him by name are influenced by his vitalist ideas about biodynamic agriculture and organic food supply. It is hard to imagine that educated people today would accept Steiner's belief in Atlantis based on clairvoyance, or his belief in other lost continents.

Two of the most zealous Pan-German and racist anti-Semitic cults were Ariosophy and the Thule Society (Goodrick-Clarke 1992, 11–16). Jeffrey Goldstein explores the mystical cultist origins of Nazism (Goldstein 1979). Though he finds the theosophy of Helena Blavatsky to have all the racist, anti-Semitic nonsense of the Nazis including the belief in Atlantis and the superiority of the Aryans that was later propagated by Rosenberg, Goldstein also recognizes that theosophy never identified the Aryans with the Germans. Although they were strongly opposed to race mixing, theosophists never advocated violence leaving a "gap" between theosophy and the Nazis (Goldstein 1979, 62). He proposes that Ariosophy "helped to bridge this gap" and it was the "avenue through which Theosophy influenced Nazi ideology" (Goldstein 1979, 62). "The resulting mixture of occultism, racism, anti-Semitism, and *voekisch* nationalism was quite volatile, one which offered a lofty 'spiritual' ideal as a justification for practices which eventually led to genocide" (Goldstein 1979, 65). The final step in the "relationship between occultism and the Third Reich" was the Thule Society, which had "direct and indirect links with all the groups in Munich which helped the Nazi Party" (Goldstein 1979, 65, 69). Goldstein is careful not to claim too much but wants the "ideological similarity and the historical connection between certain elements in these groups and Nazism" to be recognized (Goldstein 1979, 73).

The Nazis, the Occult, and Deep Ecology

Like Jonathan Olsen, Nicholas Goodrick-Clarke draws the line between the Nazi ideology and contemporary right-wing and neo-Nazi green or ecology groups, as well as a variety of New Age beliefs. Goodrick-Clarke sees a Hitler-loving mystic by the name of Savitri Devi (nee Maximiani Portas, 1905–1982) as the bridge between neo-Nazism and the New Age (Goodrick-Clarke 1998, 3). Her book, *Impeachment of Man* (written in 1945, first published in 1959, and reprinted in 1991) has "only coded references to her idol Hitler and

National Socialism"—"healthy race consciousness" being the code to appeal to "new audiences interested in mysticism and the occult, Green issues, vegetarianism, and the New Age" (Goodrick-Clarke 1998, 92, 106). The Nazi ideas she expresses "are rooted in misanthropic cult of nature and animal worship," which have become a staple of extreme (right-wing or left-wing) greens and others who view "people as a plague" upon the earth (Goodrick-Clarke 1998, 92–93, 106). She rails against man-centered beliefs, exalts the rights of animals, and with the greens and deep ecologists, she bemoans the "vicious, erosive encroachment of mankind upon the natural realm" (Goodrick-Clarke 1998, 106). In other words, excise or ignore her coded Nazi references and her work would be little different from any number of extremist green tracts and monographs and apparently has acquired an audience among them (Goodrick-Clarke 1998, 92–93). Like modern cultists, Devi is replete with vitalist terminology. She refers to Akhnaton, the ancient Egyptian pharaoh about whom she authored a book, as having "'ka' or soul, the energetic principle at the root of all existence" she wished us to worship, the "subtle Essence of Life—Energy" (Devi 1991, chapter 3). In modern cultism, using a borrowed vitalist term from another culture—*chi* or *qi* from China, *prana* from India, or *ki* from Japan—adds cache to cultism and helps make one feel superior to neighbors and friends.

Goodrick-Clarke finds that the "American movement of Deep Ecology betrays an uneasy resemblance to Savitri Devi's biocentric vision. Its precepts of community and cooperation are belied by romantic irrationalism and the assertion that all nature is equal" (Goodrick-Clarke 1998, 227). There is also a strain of racism in the American deep ecology and antihumanist opposition to immigration. Edward Abbey wanted "American citizens to consider calling a halt to the mass influx of ever more millions of hungry, ignorant, unskilled and genetically impoverished people" (quoted in Bookchin 1995, 108). Somehow humanism is now viewed as an act of "arrogance" by an author who advocates the "Noah Principle" and worries about the smallpox virus becoming an endangered species (Ehrenfeld 1978, 207–11).

The Nazis and Cosmic Occult Lunacy

One of the most preposterous of the cult beliefs was Hanns Hoebiger's cosmic ice theory in which Himmler and Hitler were "interested" (Webb 1976, 313, 325–33; Spielvogel and Redles 1986). Allegedly, Himmler became Werner Hisenberg's protector because he believed that after the war, Hisenberg would be useful in proving the cosmic ice theory to be correct (Bronowski 1965, 44). The cosmos consists of hot metallic stars, light gases of hydrogen and oxygen, and huge blocks of ice. When a block of cosmic ice falls into a star, a giant explosion follows sending off fragments of matter, some of which are captured by Earth as a series of moons that fell to Earth. The latest moon was captured by Earth only thirteen thousand years ago (Webb 1976, 326). Along with this were theories of the Earth at one time being completely covered with ice, which tied in with the theories of the German "race" having arisen in extreme cold, which made them strong.

There was also a hollow Earth theory that was considered by the Nazis and rejected (Webb 1976, 325). Nevertheless, the idea of an inner Earth was "cultivated" by the Thule Society, which was a major force in the formation of the Nazi ideology and remains potent with contemporary neo-Nazis (Goodrick-Clarke 1998, 211). Some versions actually had humans living on the inside curvature of the hollow Earth with a central core sun that revolved and had a light and dark side. Hollow Earth theories picked up neo-Nazi followers in the post–World War II period with the belief that fleeing Nazis had gone to Antarctica where they had a military base and may have entered the Earth from where they were planning to return with their flying saucers (Goodrick-Clarke 2002, 152–72; Goodrick-Clarke 1998, 3, 211–12, 227).

Many of us grew up with the popular idea that the Nazis were evil and mad but they were also rational, an interpretation that made plausible the Adorno-Horkheimer-postmodernist thesis blaming rational thought and the Enlightenment for the Nazis. These ideas were reenforced by the myths about the Nazis' technological feats such as the flying saucers. If in fact, apart from all the evil and lunacy that we conventionally associate with them, the Nazis held to an array of weird occult ideas based on rejected knowledge, then any thesis blaming modern accepted scientific knowledge for the evil they perpetuated simply falls apart. One wonders how they were able to govern a modern state for so long even though they led it to rack and ruin. Obviously, not all those in power operated in terms of these ideas. Further, we sometimes give the state too much credit or blame for how well a society performs, even in a case such as Nazi Germany, which attempted to be all pervasive. Germany was a relatively new state with many principalities still having titular rulers until the end of World War I when a large number of German speakers were scattered in other countries and may not have had the long-established mechanisms for central state control that other countries had (Goodrick-Clarke 1992, 3, 8).

In democracies such as the United States, those who come to power with a reform agenda often complain of the "bureaucratic inertia" that undermines their reform agenda. Possibly, one can say the same for Germany in the 1930s, an advanced industrial country in which large segments of the economy and society continued to function as it previously had. In other words, the sin of the professionals and other modernists in Germany in the 1930s was not for bringing Nazism into being but for not doing enough to stop it. One can imagine at the time that many of those who made things work simply assumed that the Nazis would soon pass from the scene and that their job was to keep things going until they did. Had the Nazi regime actually lost power before its worst excesses, such individuals would have been praised. At what point keeping things going made them partners in the evil that followed is easier to say in retrospect than at the time. Unfortunately this sin was so widely shared that one cannot single out a particular group for condemnation to the exclusion of others, nor does any group escape blame. Was there any group in Germany, other than the victims, that did not have large numbers who in one way or another acquiesced to Nazi rule or even participated in its worst excesses?

Ideas and the Responsibility of Those Who Hold Them

Ideas cannot be entirely responsible for those who hold them and it is a cheap trick to try to link ideas to the Nazis since it is impossible in most of our minds to separate anything that the Nazis did, however minor or trivial, from the great evil that they embodied. Pinning the Nazi (or Communist) label on someone or some ideas is simply an attempt to cease further critical thought and to close out inquiry on issues that should remain open. Conversely, because of the inextricable link with such great evil, is it thus impossible to say anything about regimes such as the Nazis except to analyze the evil itself?

Somehow we have to navigate between indiscriminate labeling and leaving a huge area of history as terra incognito for all but certain approved topics. Having said this, it also has to be recognized that ideas have a history and atavisms like biological creatures, which are not always easily discarded. The underlying fanaticism and cultic beliefs that produced evil also favored acceptance of ideas about biodynamic agriculture and homeopathic medicine. Nor can beliefs be entirely extricated from the absurd ideas that have historically been associated with them and from which they are largely derived. This is true when a subject comes up such as Hitler being a vegetarian. Instead of simply saying, so what? and moving on, true believers will simply deny it. If after World War II, civil engineers had learned that the Nazis had better technical ways to build highways, most would have eagerly adopted them since they could separate the technology and engineering of road construction from Nazi ideology. A Nazi connection is troubling to the organic advocates because there is no science that can be separated from ideology. If organic agriculture does produce safer and more nutritious food, why be surprised that those who wished to create a "master race" would advocate agricultural production suitable to this end? We are all products of culture and the ideas we hold often carry unwanted historical baggage. If an idea or body of ideas is worthy and defensible, then these unwanted atavisms can be shed. Denying the existence of an atavism won't work. Before unwanted historical baggage can be eliminated from a body of ideas, those who hold these ideas have to recognize their existence. I hope, by showing the counterexample, to stop the careless use of the Nazi designation to describe that with which one disagrees and turn the discourse and debate to the substantive issues.

We still have a right to be more than a bit uneasy when we hear the green organic agriculture proponents and devotees of homeopathy proclaiming that their ideas are becoming mainstream, even though it is highly unlikely that this will happen. It is helpful to remember the last time rejected knowledge came to power. Given lack of respect for democratic institutions, a willingness to trample on the rights of others, and intolerance of difference of opinion when they are a minority, there is reason to be concerned about the fate of civil liberties should they achieve power. We don't fear another holocaust or World War III, but this does not mean that we would not pay an enormous price if rejected knowledge became the basis for policy formation.

As is the case with homeopathic dilutions, possibly the greatest harm that is done by the greens is preventing life-enhancing and life-saving advances in science and technology from being implemented for human betterment. As the gains in life expectancy over the last two centuries have shown us, using the best in science and technology can have enormous benefit to humanity, so not using it can cause great harm, not in lives wantonly taken but in lives not saved. No better example can be found than the 2002 famine in southern Africa where drought, famine, disease, and death stalked the land. The NGO false fears and anti-GM food campaigns based on rejected knowledge have greatly hindered the relief efforts undertaken by the United States government, which supplies over 60 percent of the food provided by the World Food Program. There was simply no way that the donated maize could be certified as GM-free nor was there any food-safety reason why it should be. This difficulty was compounded by the European Union, which has used the NGO scare campaign to require GM or GM-free labeling of all grains as a means for protectionism for its agriculture, which it subsidizes at the rate of over a billion dollars a day. African leaders in the region feared that some of the donated grain would end up being planted, making future exports to Europe difficult to certify as GM-free. At no time did we hear any of the anti-GM food NGOs, some of whose annual budgets are well in excess of a hundred million dollars, offer to provide food aid. The issue was finally resolved for some of the African countries by arranging to have the maize milled before it was delivered so it could not be planted. One does not know the toll in human misery and human life resulting from the delays in delivering the food aid or from the added costs involved, but they are unlikely to be trivial. The continuing cost that the minions of rejected knowledge impose upon society are substantial.

Rejected Knowledge and Verifiable Knowledge

In this chapter, as in chapter 6, the subject is the Nazis but the underlying theme is the ideas that had strong proponents in important positions of power in Nazi Germany. Some of the occult beliefs that had adherents among leading Nazis are so preposterous that it is almost as difficult to believe that anyone in the twentieth century could adhere to them as it is to believe the many monstrous things that the Nazis did. Anyone familiar with the many culture achievements of Germany in the 1920s would find it even more difficult to understand how people holding preposterous and monstrous ideas could be in positions of power in Germany in the 1930s. These occult beliefs flourished among those who scorned modern scientific knowledge in favor of various forms of rejected knowledge and romantic views of nature. Once again we have a cautionary tale as various occult forms of rejected knowledge and romantic notions about nature being benign are flourishing today. It would be wrong to draw any further analogies to the Nazi era as likely outcomes today, but nevertheless there is reason for intelligent concern and a continuing need to counter rejected knowledge with verifiable knowledge based on the scientific method and free and open inquiry.

CHAPTER 8

Vitalism, the Organic, and the Precautionary Principle

W e now have the "precautionary principle" with dozens if not hundreds of definitions for it. One clever expression states that "absence of evidence of harm is not evidence of absence of harm." Not mentioned is the fact that absence of evidence of harm is sometimes the only evidence possible that there is no harm. As has been repeatedly maintained by scientists, they cannot prove there is absolutely no possibility of harm, now or in the future; all they can show is that the best scientific testing can find no evidence of harm and nothing in our current scientific knowledge gives us any reason to expect to find harm by continued testing.

Green opponents of genetically modified food will not accept as evidence that the number of people who have eaten genetically modified food now number into the hundreds of millions without there being a single verified instance of even the slightest hint of harm to anybody. There are a number of peer-reviewed articles in leading journals that have examined the safety of GM crops and could find no dangers (for example see Chassy 2002 as well as the numerous peer-reviewed and panel studies that he cites). We cannot ultimately prove the safety of anything we do. Certainly, by the criteria demanded, no novelty could ever be permitted.

When it serves the cause of being opposed to a technology, absence of evidence of harm can become evidence of harm even if the likelihood of harm is deemed unlikely and all the scenarios by which harm could be brought about are highly implausible, if not impossible. This is precisely what a distinguished scientist serving as an adviser to the BBC said concerning a two-part drama, "Fields of Gold," on a global catastrophe resulting from a genetically engineered crop (Derbyshire 2002; Henderson 2002; Shouse 2002; Wells 2002). That did not deter the BBC from going forth with the highly publicized two-part drama that made a close to impossible event seem very real and highly likely and would have reinforced all the phobias of those who view such a catastrophe as a certainty. That the scientist described it as being ridiculous, a ludicrous piece of alarmist science fiction, mattered little. There was sufficient license to go forward with the project as long as the scientist did not state categorically and unequivocally that the scenario was absolutely impossible. Unfortunately, the magic words "absolutely impossible" were precisely the words that no reputable scientist would even whisper. Absolutist assertions are the mainsprings of ideology, not science. Attempts to label the criticism of the science advisor and other scientists who added their voices as being part of an "ugly conspiracy" and smear campaign orchestrated by the "pharmaceutical giants" were ludicrous and implausible. The science adviser, a distinguished Cambridge scientist and an expert on plant genetics, was a former member of Greenpeace and self-described as a green socialist (Rusbridger 2002).

A clear pattern has emerged. Given that modern culture still has some reserve of respect for science in spite of Luddites and postmodernists, the critics of a technology will seek to simulate science in their attack against it, including an occasional but rare peer-reviewed article. Modern science is presumed to be fundamentally erroneous and destructive except when it can be used to support the critics' position. When leading scientists counter an article or fictionalized program on scientific grounds, there is an inevitable response charging a conspiracy orchestrated by malevolent multinationals with the scientific issues all but ignored and in some cases totally ignored, as was the case of the BBC drama. The response by one of the authors, the novelist Ronan Bennett, of being a victim of "an ugly conspiracy by those with a vested interest in discrediting it and personal grudges to settle," in no way responds to the substantive issues raised by the scientists who made no personal attacks against those involved with the TV drama (Hume 2002). The rhetorical response in this instance is virtually identical to that in every other case where an antitransgenic claim is challenged on scientific grounds, the terms *smear, conspiracy,* and *pharmaceutical multi-nationals* are immediately raised, which might justifiably lead one to conclude that the critics have no scientific basis for the opposition to the technology and are aware of that fact.

Given the hysteria that certain groups have effectively stirred up over genetically modified or transgenic crops, such a presentation was simply irresponsible and could not be justified on the basis of literary license. The BBC claimed that "Fields of Gold . . . is a fictional drama which does not purport to be a documentary," and compared the "very real fear" that people have about transgenic food "the way nuclear power terrified people" (Hume 2002). The comparison is more apt than the BBC realized because both of these phobias simply lack any basis in fact and they are perpetuated by groups whose continued survival depends upon frightening people about modern science and technology. One doubts that the BBC or any other major television organization would use literary license to justify producing and airing a drama in which some fanatical greens are able to precipitate a global nuclear conflict or some other global catastrophe.

Antiscience, antitechnology programs are not without their adverse consequences. Given the lives that have been improved and prolonged by organ transplants, most people would acknowledge that the practice is "life affirming and marvelous." One author reminds us of another BBC broadcast "Panorama," which in the 1980s ran a program, "Are Transplant Donors Really Dead?" "Although this was a farrago of inaccuracy, organ donation was reduced for a decade owing to change in public sentiment" (Gray 2002). Who involved with the program will take responsibility for the lives that could have been saved but weren't because donor reluctance resulted from this program? Those who expect scientists to examine the short- and long-term consequences of their actions and be responsible for them, as they should be, are themselves singularly unable to accept any responsibility for their actions.

Frankenfears and the Internet

In the war of words over transgenic crops, opponents have successfully coined a number of pejorative terms—Frankenfoods or Frankenstein foods, genetic pollution, contamination among many others—

most of which they have successfully gotten the media to use. Given that there is no substantive science to support the use of these pejoratives, it would be fair to call them Frankenfears.

Over the last decade, there has been one scare after another concerning transgenic crops; many seem strategically timed to precede an important international or regional meeting. No sooner is one massively refuted, than another conveniently arises. In cyberspace, the world of myth making never dies and seems to live on and on. Since there is no cyberspace in reality, only a series of interconnected servers, this means that these myths are being deliberately perpetuated by those who are writing articles and keeping them on their servers. Of the many scare scenarios that have been promulgated, one of the most bizarre was the submission to the New Zealand Royal Commission on Genetic Modification in February 2001, which claimed that the release of the genetically engineered bacterium *Klebsiella planticola* "would result in the death of all terrestrial plants" by entering the root system of plants causing the production of more than sufficient toxins to kill them. The following sentence in the executive summary modifies this assertion slightly by claiming there would be "devastating impact on humans" because we would have "lost corn, wheat, barley, vegetable crops, trees, bushes, etc., conceivably all terrestrial plants." The headlines in the New Zealand press—" 'GM bacteria could kill all life' US expert"—made clear how this submission was interpreted, an interpretation that was not denied by the greens who sponsored the submission.

The counterattack by the scientists who investigated the issue was careful, reasoned, and as devastating to the credibility of the scientist who gave the testimony as the bacterium was alleged to be to humans. In spite of the usual claims of a smear by a Greenpeace scientist, the scientist and the Green Party had to back down from the submission in a letter to the commission in which they "apologized for submitting false claims about the ecological impact of genetically modified organisms" (Fletcher 2001). A reputable scientific publication called it a "debacle" and another "example of the hijacking of scientific research for political ends" (Fletcher 2001).

One would think that this would end the matter but unfortunately, it has not. Anyone new to this issue who goes to a search engine like Google and types in *Klebsiella planticola* or *Klebsiella bacterium*, will get a huge list of sites. Topping this list are a number of sites with articles elevating the miscreant scientist, Dr. Elaine Ingham, to a mythic heroic figure for having saved life on earth by preventing the release of the genetically modified bacterium upon the earth. Two examples of article titles: (1) "Klebsiella planticola—The Gene-Altered Monster That Almost Got Away," (2) "A Biological Apocalypse Averted," a posted excerpt from a book that has a diet that can "Save Your Life and the World." It is difficult to decide whether it is worse to believe whether those who continue to write articles and post them are simply without any sense of shame and are propagating ideas that they know to be false, or whether they are so zealously committed to an ideology and so blind to reality that they believe them to be true.

There is a posted defense by Dr. Ingham arguing that she never claimed "all life" on earth would be killed. Technically this is true from a cosmic or deep ecology perspective but the differences between what she said and what she claims not to have said are important, but from the "species centric" view of humans, either of the two is a catastrophe of such immense magnitude as to be almost beyond comprehension and something none of us would knowingly act to bring about. Dr. Elaine Ingham who is the president and director of research, Soil Foodweb Inc., Corvallis, Oregon, "serves on the boards of a number of sustainable-oriented organizations, and speaks to groups around the world about the Soil Foodweb and how to grow plants without the use of pesticides or inorganic fertilizers" (Ingham 2003).

At least two different postings of Ingham's defense came up along with the praise postings prior to anything indicating against what charges she is defending herself. Going down through the list, there was one article by Liz Fletcher in *Nature Biotechnology* that gave a good account of the "debacle" (Fletcher 2001). Further search down the listings would have undoubtedly turned up the posting by scientists refuting Ingham claims. What is clear is what nobody denies, namely that the greens have mastered the art of propaganda, including the use of the internet and knowing how to design web pages (with links to ideological soul mates), so that they come up first in any literature search. An honest inquirer, such as a reporter who heard the issue raised for the first time and went online to learn more, would have to make a considerable extra effort to obtain a balanced account. Those on deadline, might not have the time to do so allowing the greens to score with another rendering biased by their having dominated the online sources. (For superb articles on the use of propaganda on biotechnology issues, see Logan; Palfreman; Hines; Marks and Kalaitzandonakes; Blaine and Powell, AgBioForum 4(3,4), 2001; Santerre and Machtmes 2002).

Precautionary Principle: The Double Standard

Given that there are dozens or possibly hundreds of definitions of the "precautionary principle," with most revolving around some form of absence of evidence of harm not being evidence of absence of harm, it means that, lacking any evidence upon which to make an intelligent judgment, we will need an anointed class to define what is safe and what is not. Absence of evidence is about the only evidence that we can have, which means anything we do is potentially harmful. Thus, we will have to call upon a green priesthood, the contemporary entrepreneurs of salvation, who have the mystic powers to divine what absence of evidence is really evidence and what is not. And entrepreneurs they are. We must never forget that the pecuniary motive is as an operative for a chain of "natural food" stores or for an environmental organization with annual income and expenditures in the hundreds of millions of dollars, as it is for any other organization.

The new green priesthood is not without its double standard. The authors of a widely touted study purporting to show lower levels of pesticide in organic produce, nevertheless admit to there being a "significant gap in this qualitative risk comparison, related to the possible contribution to total risk of residues of natural pesticides" (Baker et al. 2002, 442). In other words, one tests for lower levels of pesticide residue by testing both conventional and organic produce for the pesticides used in conventional agriculture but not those used in organic agriculture. One could have reasonably expected their finding without the need for empirical investigation. *Consumer Reports*, which has a history of comparing "pesticide residues" on conventional and organic produce by omitting a reference to "natural pesticides," touted this report with the headline "Organic is lower in pesticides. Honest" (CR 2002, 6). A good rule of thumb would be that if someone follows an assertion with the exclamation "honest," it could mean they are deliberately fudging the truth, as no mention is made of natural pesticides (see USDA 2002 for the approved list of organic pesticides; see also DeGregori 2001, 85–86). Environmental groups made similar claims following the report's release.

The authors try a number of stratagems to close this "significant gap" in their data including surveys of organic farmers and indirect evidence that the botanical and other pesticides used by organic farmers are either not harmful or not present in doses to pose a threat to human health (Baker et al. 2002, 443). E. Groth, one of the authors of the study, summed up the article's findings on this point as follows: "At present there are no good residue data on the botanicals and other natural pesticides, and some of those substances definitely should be more fully evaluated for potential toxic effects" (CUPR 2002).

Groth adds: "There is now no objective evidence of a problem with residues of natural pesticides, whereas health risks associated with residues of conventional pesticides in foods are well established and the focus of substantial regulatory efforts" (CUPR 2002).

In the Consumers Union press release, it admits to not testing for the natural pesticides but still claims that they are safe. No testing, therefore no evidence. It is funny that lack of evidence of harm for genetically modified foods is evidence of harm but lack of evidence of harm for natural pesticides is evidence of safety. There is more than a bit of ideology operating here. At least they are saying that "some of those substances definitely should be more fully evaluated for potential toxic effects." If nobody is testing for them, how can they be "more fully evaluated for potential toxic effects"? Should this be a task for Consumers Union/Consumer Reports? If it is an "interesting question," may I inquire as to why it is not being asked and why research for answers has not been pursued?

Given the claim that the natural pesticides are harmless, maybe we should identify some of them. In the United States, the oil and sulfur used in organic agriculture are by far the most heavily used pesticides. Sulfur is a concern because of its persistence in the environment; other organic pesticides, like rotenone or pyrethrum, are carcinogenic or otherwise toxic to humans (Avery 2002b). Under the rules of the Soil Association ("the ayatollahs of the organic movement" in the United Kingdom), "organic farmers are permitted to use a list of products all of which are manufactured in factories and are not made by little old ladies boiling up seaweed on a Cornish clifftop" (Walston 2002). "Organic farmers are today allowed to use 14 different fungicides and eight different insecticides. Among the latter are the potassium salts of fatty acids, often called soft soap. . . . Note the following statement: 'Harmful to fish and other aquatic life. Do not contaminate waters or ditches'" (Walston 2002).

One need not engage in any arcane semantic discussion to see the outright falsity of continuing to claim that no pesticides are used. One does not put the agronomic equivalent of a condom on an organic plant or otherwise hermetically seal it from the environment. One puts substances on plants that are toxic to (or otherwise seeks to protect plants from) insects, rodents, birds, viruses, bacteria, or fungi. The purpose of plant protection is to kill microorganisms and to either deter or kill other pests. If it kills anything harmful to the plant, it is in some sense a pesticide in the full meaning of that term. In other words, it is a toxin or poison, which is lethal to some living organism. We can argue the relative merits of the pesticides used in organic agriculture versus those used in conventional agriculture as to whether they are also toxic to humans, the natural predators of the pests, other nontarget species, or the environment in general. But one can not honestly say "pesticide-free" when it is not true. Admitting this is intolerable to the true believers since their use of the language clearly implies that being a toxin or a poison or pesticide is an absolute and must be poisonous to all living creatures.

When one sees reality in only black and white terms and can think and act only in terms of absolutes, then one has to be continually in denial concerning any nuances that conflict with one's personal reality. Thus, they have to continually repeat the mantra "pesticide-free" when it is simply not true. Without question, modern synthetic pesticides replaced far more toxic natural pesticides than those in use such as arsenic compounds. A good argument could be made today that many of our synthetic pesticides are less toxic than those used in organic agriculture. With the advent of crop protection built into transgenic crops, which the organic folk are resolutely against, there will be little question as to which crop, conventional or organic, has the fewest toxins, either applied by the farmer or produced by the plant. With transgenics, conventional farmers will be able to produce a crop as close to being truly pesticide-free (the only pesticide possibly being a gene that expresses a protein toxic only to specific pests) as has ever been done by humans.

Vitalism and Virtue

What is unstated but very clear is that believers in various contemporary forms of vitalism also believe that there is a certain higher virtue and morality to these beliefs and practices and a corruption and even evil to those that differ. Many of these issues of vitalism have become consumer issues in recent years. Hamilton offers a clarifying vision in evaluating consumer products in terms of their instrumental efficacy and not in terms of some alleged "mystic potency" (Hamilton

1999). Previously, theologians claimed a mastery of the unknown and unknowable. Today it is the greens who claim to have an unchallengeable claim on all the gaps in our knowledge about dangers not provable and expertise on that about which we are ignorant. To argue that "the gap in knowledge which still confronts the seeker must be filled, not by patient inquiry but by intuition or revelation, is simply to give ignorance a gratuitous and preposterous dignity" (Mencken 1930, 307).

Consumers pay a premium for that which is natural, but who has clearly defined what is natural and what is not? Consumers will go to their health food store and pay a premium price for produce because it is natural and organic and grown without chemicals even though they may have been grown using toxic compounds of copper, sulfur, or arsenic. Organic pesticides include pyrethrum, a carcinogen, and rotenone, which has been shown to be a precursor of Parkinson's disease. Rotenone has long been the poison of choice for killing fish both by American Indians from pre-Columbian times and by the U.S. Fish and Wildlife Service (Potter 2002). Our consumers will then wheel their carts over and pay a premium price for all natural amino acids. These are made in huge vats in Japanese chemical factories using genetically engineered bacteria with feedstock that has to be filtered by multiple charcoal processes and reverse osmosis to create what largely does not exist in nature, a free-standing amino acid. We get our amino acids from complex proteins. They are active compounds that have largely not been tested for either their efficacy or harm in the doses in which they are taken.

One group of researchers chose to do a twenty-one-year project on 1.4 hectares of land comparing conventional, organic and biodynamic agriculture and was able to publish its findings in one of the most prestigious scientific journals (Mader et al. 2002; Stokstad 2002). By any reasonable standard, a labor-intensive study by advocates of organic agriculture on 1.4 hectares of land, even a twenty-one-year study, would provide very little of value, even if the findings had been anything more than the meager results that were claimed. The article was widely touted. Editors of the very top journals in science know that an article favorable to organic agriculture or some other green cause or one that indicates possible harm from a technology that the greens oppose will get instant massive publicity no matter how weak the evidence may be. The publication will often alert the media in advance about a forthcoming piece. This may be good for the publication but it harms science. The desire to be balanced and open to contrary ideas may lead to publishing pieces that might not otherwise have made the cut. Given that the journal frequently knows in advance that a piece will draw publicity, it should alert the media to other scientists with a different take on the issue so the media that have a green bias do not have love feast interviews with the senior author or at least have the opportunity to examine a different view should they so decide.

Conservation of Wildlife but Not Human Life

Before the war, Himmler had great plans for creating an "enormous wildlife preserve" out of Poland, which were never carried out (Teresi 2000). The post–World War II period has seen an enormous increase in areas dedicated to wildlife conservation. "From less than 1,000 protected areas in 1950, the count grew to 3,500 in 1985 before ballooning to 29,000 today" (Geisler 2002). "Nearly twenty-nine thousand protected areas now shield some 2.1 billion acres of land from a series of residential and economic uses. These territories compose 6.4 percent of the earth's land, equal to about half the world's croplands, and are roughly the size of the continental United States plus half of Alaska. Most of this protection is recent"(Geisler 2002). One conservationist who "flies from project to project" has decided that "throwing the tribals out of a park, he says, is the only means to save the wilderness." Consequently, "relocating people who live in these protected areas is the single most important step towards conservation." The tribals "compulsively hunt for food" and "compete with tigers for prey" and must be moved (quoted in Guha 1998, 275–76; see also Guha 1989, 1997). Too bad they can't obtain food from supermarkets like our intrepid conservationist. Another finds the locals unworthy to protect the environment; they are not as informed about it as its visiting protector. "Virtually all of the present-day occupants of western Mesoamerican pastures, fields and degraded forests are deaf, blind and mute to fragments of the rich biological and cultural heritage that still occupies the shelves of the unused and unappreciated library in which they reside" (quoted in Guha 1998, 274).

It is still too often the case that humans occupying the land in low-income tropical regions, such as parts of Africa, are considered less important than the flora and fauna that are to be protected. "In 1985, Africa had 443 publicly protected areas encompassing 217 million acres of land. Facing international pressure, virtually all African countries have since increased their protected land base" (Geisler 2002). There are 380 million acres of protected land in Africa including fourteen African countries that have more land in conservation than under cultivation. There are estimates of the number of humans that have been displaced; they range from nine hundred thousand to 14.4 million. Even the lower bound number is astounding and it is likely that the real figure is considerably higher. The displaced have been called "endangered humans" and "invisible refugees" (Geisler 2002, 80–81). "Nearly half of the planet's most species-rich areas contain human populations suffering severe economic disadvantages. The tropics, where biodiversity flourishes most, are home to nearly 60 percent of the world's most destitute people" (Geisler 2002, 81).

Charles Geisler recognizes the need for conservation and "green consciousness." He still has powerful comments about the injustice in the ways that it is most often carried out. "While global conservation does not cause poverty, neither should it exacerbate poverty. The poor should not be asked to disproportionately subsidize the expansion of conservation. They, too, must have voice and choice. If conservationists are to retain the mantle of justice, they must find alternatives to involuntary and uncompensated human displacement" (Geisler 2002, 81).

When all else fails and there is no evidence of harm, opponents of transgenic food crops have invoked the precautionary principle. The greater the imagined fear, the greater the justification for opposing a new technology no matter what the facts of the case may be. Vitalist principles cannot be proved and to the believers need not be. Organic agriculture has from the beginning been based on vitalist principles, so its supporters seek to use science to try to support what they already assumed to be the truth rather than using open-ended scientific inquiry to ascertain what the truth actually is. Their conclusions, then, should not come as a surprise to anyone.

We must never forget humans possess a history because we are not "bound" to nature, instinct, or "biological processes alone" (Ferry 1995, 38). "*All valorization, including that of nature, is the deed of man and that, consequently, all normative ethic is in some sense humanist and anthropocentrist* (Ferry 1995, 131). This is how he escapes natural cycles, how he attains the realm of culture and the sphere of mortality, which presupposes living in accordance with laws and not just with nature (Ferry 1995, XXVIII).

CHAPTER 9

Feeding Six Billion People

Entering the twentieth century, the world's population was about 1.6 billion people after a century of unprecedented population growth fueled by roughly one percent per year growth rates, which were also unprecedented in human history. Sometime early in the nineteenth century, the world's population had, for the first time, crossed the one billion mark and kept right on growing. Those of us today who complain about unfettered population growth in Africa and other poor areas must never forget that in 1900, the population growth from one half billion, tripling to over one and a half billion during the course of the previous four centuries was driven largely by European and, to some extent, Chinese population growth. In areas like Africa, population growth was much slower or even remained relatively stationary.

There is evidence of a decline in fertility by the 1830s in France followed by other western European countries as the century progressed, but it was not until late in the century that declining fertility rates were translated into birth rates falling faster than death rates, leading to declining rates of population growth. The forces of population growth continued in overseas European populations in North America and colonies and former colonies around the world. By 1950, the forces of more rapid population growth had been unleashed globally as population growth rates would accelerate, reaching a peak of about 2.1 percent in the late 1960s and early 1970s before beginning a slow decline.

Feeding 1.6 billion people in 1900 was an extraordinary achievement considering that an increasing proportion of the fastest growing populations were better fed than any comparable populations in the past. The transformations in science and technology in Europe were being realized in manufacturing in what we now call the Industrial Revolution. The work of Liebig and others in chemistry and agronomy were providing greater yields in agriculture as a result of plant breeding and more effective fertilization. New lands were being opened to cultivation and cattle and sheep raising, and railroads were bringing the products to the ports. Improved sailboats and steamships, some refrigerated, were bringing diverse foodstuffs to the European markets.

Despite the optimism of the time, a limits-to-growth theory could have raised serious questions not only about whether the growth in food supply could continue but whether it could be sustained at its then current level. Europeans and North Americans were literally mining guano and nitrates in the rest of the world to provide the nutrients for their agriculture and food production. These were clearly exhaustible resources that were becoming increasingly scarce. The frontier in the United States had officially been declared closed; a new nation was complete. Food production was actually growing more rapidly than population with improved nutrition being a factor in the increases in life expectancies and decreased infant and child mortality, that were the sole cause of the increase in population growth rates.

The knowledge about plant nutrition that greatly contributed to the nineteenth century growth in food production, particularly Liebig's Law of Minimum, that growth of an organism was limited by the least-available nutrient, made clear the potential dangers that were just over the horizon. For agriculture, the potentially limiting nutrient was nitrogen in a form that was usable by plants. The guano was being mined to exhaustion and the sources of mineral nitrates were being depleted. Given the knowledge of the time, it is a wonder that pessimism about the future of humankind did not prevail. Had someone forecast that in a single century, the world's population would top six billion, the first response would have been that it cannot happen. This would immediately have been followed by forecasts of the specters of famine, disease, and death stalking the planet if it did happen, with the fabled Four Horsemen of the Apocalypse becoming a thundering herd of death-dealing, hooded cavalry. But population growth did happen and the century ended with a world population better fed than ever with a larger population living in relative affluence at the end of the century than there were people at the beginning of it. This is a truly remarkable story. And nobody has told this story better than Vaclav Smil in an impressive series of books and articles, most recently, *Feeding the World* and *Enriching the Earth* which deals with nitrogen, the major limiting nutrient for life on earth as we know it and therefore for the plant and animal life that sustains us (Smil 2000, 2001).

"In 1900 the global count stood at 1,625 billion after a century of growth averaging 0.5% a year. Twentieth-century growth averaged 1.3% and as the year 2000 began the global population just surpassed 6 billion people" (Smil 2001, 199). Smil sums up the amazing span of 100 years as follows: "While the twentieth-century population total has grown 3.7 fold, the global harvest of stable cereal crops, as well as the total production of crop derived food energy, has expanded sevenfold. As a result, never before have so many people—be it in absolute or relative terms—enjoyed such an abundant supply of food" (Smil 2001, 199).

Nitrogen Use and Organic Agriculture

Before turning to twentieth century growth in food production, I need to provide some historical perspective for the issues involved in this remarkable story. The history of nitrogen use is interesting, important, and worth telling for its own sake. It takes on added importance in terms of the set of beliefs about so-called organic agriculture, the early formulation of which we covered in chapter 8. In origin, organic or "biodynamic" agriculture was part of the continuing advocacy for the exclusive use of organic matter for plant nutrition. Opposition to the use of synthetic fertilizer beginning in the 1920s was a continuation of the vitalist dissent from Woehler and Liebig and the foundation of organic chemistry a century earlier. Cut away the contemporary rhetoric from the opposition to synthetic fertilizer and one finds a vitalist argument for the superiority of manure. The widespread use of synthetic pesticides arose well after the beginnings of the biodynamic movement in agriculture and could not have been the cause of it. Paeans in praise of manure are uttered by its advocates and it is claimed that organic agriculture can be conducted without fertilizer and that such practices can be used to feed the world. The beliefs about the alleged superiority of organic produce among the general public are generally quite vague, though even here among the marginal believers, there is a hint of vitalism. There almost has to be a vitalist core as there is no scientifically verifiable evidence conferring any nutritional benefit to organic produce. There is even a Vegan Organic Network (www.veganvillage.co.uk/vohan/), which with true vegan consistency opposes the use of animal manure (since vegans oppose raising animals for food). They believe that green manuring and recycling human waste can provide sufficient plant nutrient for all our agricultural needs. Others oppose the use of human waste from treatment plants, as it is likely to contain heavy metals.

It is not always clear what precisely is meant by claims for organic agriculture, but if we are to understand the critical issues in modern agriculture, we have to have a detailed, basic understanding of the role of nitrogen in agriculture, the critical role of synthetic nitrogen fertilizer for twentieth- and twenty-first-century agriculture and nitrogen as a constituent of the amino acids that form the proteins that nourish us. The misunderstandings on these points make the story of nitrogen worth telling.

Nitrogen and Amino Acids

Long before the looming twentieth century crisis of nitrogen deficiency for agriculture, life on earth faced a potential nitrogen crisis. Nitrogen is abundant in the earth's atmosphere but it is not in a form (called "fixed") usable by life to create the amino acids that form the basis for proteins (Smith 2002). "Nitrogen is an essential element for life and is often the limiting nutrient for terrestrial ecosystems. As most nitrogen is locked in the kinetically stable form, N_2, in the earth's atmosphere, processes that can fix N_2 into biologically available forms—such as nitrate and ammonia—control the supply of nitrogen for organisms. On the early earth, nitrogen is thought to have been fixed abiotically, as nitric oxide formed during lightning discharge" (Navarro-Gonzalez 2001).

The standard definition of an amino acid is an organic compound that contains an amino group, NH_2, and a carboxylic acid group of the type COOH, various combinations and sequences of which are the basis of all proteins. "The genetic code of all organisms encodes the same 20 common amino acids" (Wang et al. 2001, 458). Without nitrogen there would be no life as we know it on earth. "The chemical invariance of the 20-amino acid building blocks of proteins is well established" (Doering et al. 2001, 501). "One of the most intriguing aspects in the evolution of the genetic code—the 61 nucleotide sense codons in the DNA that encode the amino acids of all organisms—is why the amino acid complement is limited to the magic number of 20. This number is invariant—from archaea and bacteria to eukaryotes—irrespective of their evolutionary complexity and the chemical and physical environment in which they live" (Bock 2001, 453).

Bacteria that have adapted to extreme environments do so by "shifting the proportions of the 20 amino acids in existing proteins, which carry out the same tasks regardless of where the bacteria live." In other words, "psychrophilic (cold-loving) bacteria—which grow at temperatures of 20 degrees C to below freezing point—are composed of the same 20 amino acids as those of hyperthermophilic bacteria, which exist at temperatures of 100 degrees C or more" (Bock 2001, 453). We can still speak of the twenty "naturally occurring" amino acids but scientists have found a twenty-first and twenty-second "nonstandard" amino acid (Atkins and Gesteland 2002; Hao et al. 2002; Wills and Bada 2000, 12–13; Srinivasan et al. 2002). It remains true, however, that the "great majority of nonstandard amino acids are created by chemical modifications of standard amino acids after translation" and are as unlikely to turn up in a transgenic plant as in one from conventional breeding (Atkins and Gesteland 2002). While claims are made against transgenic crops on the basis of a novel potentially allergenic protein being transcribed, no critic has yet raised the possibility of a novel amino acid emerging since the possibility is so remote. Plants produced by conventional breeding are even more likely to produce a novel protein or a greater amount of a highly toxic protein that is already present in very minute quantities than is the case for a transgenic. "Despite life's vast diversity, all creatures—from yeast to humans . . . spell out their genetic instructions using the same four DNA chemical units, known as bases, which are represented by the letters A, C, G and T" (Pollack 2001).

A is adenine, C is cytosine, G is guanine, T is thymine. Uracil (U) replaces thymine (T) in RNA molecules and nucleotides with the other three bases retaining the same name but with each adding one oxygen. Differing combinations of these letters code for different amino acids, which combined in various sequences create the proteins that "carry out most functions in a cell. . . . The genetic code, then, is a language of four letters" grouped in sets of three called codons, which provides an alphabet of 4^3 or 64 letters to make the twenty words (the standard amino acids), which are used in living organisms to make "the huge variety of sentences and paragraphs that characterize life" (Pollack 2001; see also Tanford and Reynolds 2001, 237; Bradley 2002).

The discovery of this language of life began with the determination of the double helix structure of the nucleic acid, DNA as the "self-perpetuating carrier of genetic information." From this Crick stated the famed "sequence hypothesis," namely that the "*sequence of bases* in any section of cellular DNA—perpetuated by the complementarity of the double helix—must uniquely determine the *sequence of amino acids* in a corresponding peptide chain" (Tanford and Reynolds 2001, 237). This switched the emphasis from the chemical structure to the "information content." "A new non-chemical language had to be invented just to express the idea, a metaphor based on 'language' itself—molecular information portrayed as analogous to letters of the alphabet and their ability to create meaningful words" (Tanford and Reynolds 2001, 237).

Increasingly, an understanding of the molecular basis of life and its history have become an essential component of the more applied knowledge of food production necessary to feed a population of six billion and growing. What once would have been considered abstruse but interesting knowledge about life's origins and development on earth are now inextricably linked to most all aspects of contemporary life sciences.

Earth's First Life

The earth's first life forms were heterotrophic, which means that they could not manufacture their own nutrient and therefore subsisted on a cumulated store of organic matter that had been created and was continuing to be created by lightning discharged in the earth's atmosphere. "There was abundant energy from the sun and in the ocean from sulfides" but the "invention of replication was more difficult. DNA is now (with the exception of some viruses) known as the molecule that is indispensable for replication" (Mayr 2001, 43).

As life expanded, it was using more organic material than was being created, preexisting stocks were being depleted, and "at some point the demand for fixed nitrogen exceeded the supply from abiotic sources," creating a "possible nitrogen crisis for Archaean life" (Navarro-Gonzalez 2001). As long as adequate sources of nitrogen were available for early life, there would be no survival advantage if the ability for nitrogen fixation emerged, since "biological nitrogen fixation is energetically expensive" (Navarro-Gonzalez 2001). But when demand for nitrogen exceeded the abiotic supply, whether from depletion of prebiotic sources or from the emergence of higher plants, the development of energetically expensive "metabolic pathways to fix nitrogen" could have a survival value if it arose, which it did (Navarro-Gonzalez 2001; see also Heckman et al. 2001). Fortunately for us, life evolved energetically expensive metabolic pathways to fix nitrogen, allowing life to continue to evolve higher forms. "The biochemical unity that underlies the living world makes sense only if most of the important molecular types found in organisms, that is, most of the metabolic pathways involved in the production of energy and biosynthesis or degradation of the essential building blocks already existed in very primitive organisms such as bacteria" (Jacob 1977, 1165).

In these emerging higher life forms, there were new uses for nitrogen. "Although relatively scarce, nitrogen is present in every living cell: in chlorophyll whose excitation by light energizes photosynthesis (the biosphere's most important conversion of energy); in nucleotides of nucleic acids (DNA and RNA), which store and process all genetic information; in amino acids, which make up all proteins; and in enzymes which control the chemistry of the living world" (Smil 2001, xii–xiv).

In addition, nitrogen is the "nutrient responsible for the vigorous vegetable growth, for the deep green of the leaves etc. . . . and protein content of cereal grains, the staples of mankind" (Smil 2001, xiv). "Nitrogen (N) comprises fully 16% of protein" in all living matter (Frink, Waggoner, and Ausubel 1991).

Unlike the predecessor heterotrops, plant life could secure their energy needs from the sun by photosynthesis. But they still needed help in obtaining their nitrogen needs. The atmospheric N_2 molecules "must be split into two constituent atoms before they can be incorporated into an enormous variety of organic and inorganic compounds" and for this they needed another organism, nitrogen-fixing bacteria. "There is only one group of living organisms capable of nitrogen fixation, about 100 bacterial genera, most notably Rhizobium bacteria associated with the roots of leguminous plants" (Smil 2001, xiv; also Lane 2002, 145;

for closely related *Agrobacterium tumefaciens*, Ditt, Nester, and Comai 2001).

"A tight metabolic association with rhizobial bacteria allows legumes to obtain nitrogen compounds by bacterial reduction of dinitrogen (N_2) to ammonium (NH_4^+)" (Galibert et al. 2001, 668; see also Einsle et al. 2002). Free-living nitrogen-fixing bacteria are called diazotrophs and can "sever dinitrogen's strong molecular bond at atmospheric temperature thanks to nitrogenase, specialized enzymes no other organisms carry" (Smil 2001, 19; Smil 2002, 67–68). "This capability is quite remarkable because the enzyme responsible for reducing N_2, nitrogenase, is poisoned by O_2. Thus, cyanobacteria have had to evolve complex mechanisms for protecting their nitrogenase" (Kasting and Siefert 2002, 1066). In "living cyanobacteria, NH_4^+ depletion induces the formation of a transcriptional regulator," which releases enzymes that "efficiently scavenge bioavailable N from seawater" (Anbar and Knolle 2002). "Harvesting light to produce energy and oxygen (photosynthesis) is the signature of all land plants. This ability was coopted from a precocious and ancient form of life known as cyanobacteria. Today these bacteria, as well as microscopic algae, supply oxygen to the atmosphere and churn out fixed nitrogen in Earth's vast oceans" (Kasting and Siefert 2002, 1066).

There are some lessons here applicable to human resource issues. Nitrogen in a form usable by the existing heterotrophic life was a fixed finite resource and as such was inherently exhaustible even though in this case it was being renewed by abiotic sources. "Living within limits" would have meant, at best, no evolution to higher forms, though one has difficulty imagining the mechanism for a continued expansion of life in some form to the limits of resource availability without increased death being the force that keeps resource supply and demand in balance. Though nitrogen is essential, it is not a resource unless it is in a usable form. Beyond the fixed nitrogen created by lightning, atmospheric nitrogen, roughly 80 percent of the earth's atmosphere by volume, *became* a resource when life evolved the means of using it. As I often quote and will continue do so often, "resources are not; they become" (Zimmermann 1951, 15). "Microorganisms may also have played a major role in atmosphere evolution before the rise of oxygen. Under the more dim light of a young sun cooler than today's, certain groups of anaerobic bacteria may have been pumping out large amounts of methane, thereby keeping the early climate warm and inviting. The evolution of Earth's atmosphere is linked tightly to the evolution of its biota" (Kasting and Siefert 2002, 1066).

Nitrogen and Reductionist Science

The much maligned reductionist science of chemistry, which was able to synthesize organic compounds in the nineteenth century, was able in the twentieth century to perform what may be the most important synthesis of the century, the synthesis of ammonia and its use in the creation of urea for agriculture. Smil calls this—the (Fritz) Haber (Carl) Bosch process—the "most important technical invention of the twentieth century" (Smil 2001, xiii). Humans could now take this most abundant of the atmospheric elements and convert it—manufacture it—into the most vital resource for the growing of crops and the creation of nutrients to feed humans. Work in the laboratory to find ways of reversing a process that was already known, the breakdown of ammonia, began in the late nineteenth century, was successfully carried out in the laboratory in the first decade of the twentieth century, was a functioning industrial process by World War I, and has continued to become more efficient since then (Smil 2001, 61–107).

As we have seen in previous chapters, the vitalist reaction to organic chemistry was almost immediately applied to synthetic fertilizer, and the food thereby produced was attacked as lacking some vital life force. The vitalist attack against modern science and its fruits continues to the present as does the skepticism about the "project of mastering nature" (Rifkin 2001; *The Economist* 2001; Jackson 2002; Fukuyama 2002; and McKibben 2002). Rifkin has proclaimed ours to be "The Age of Biology" and has been touting that the Left and the Right are finding in vitalist principles a "common ground in opposition to a utilitarian view of life" (Rifkin 2001). There is now an impressive array of "leading thinkers and commentators" across the political spectrum who oppose "criminalizing science" by banning "therapeutic cloning" and generally stigmatizing genetic research (Postrel 2001). Rifkin's vitalist alliance isn't the "only odd meeting of minds under way between academic trendiness and religious fundamentalism" as Matt Cartmill describes "the strange alliance against Darwin that's emerged in recent years between the forces of the religious right and the academic left" (Cartmill 1993 quoted in Olson 1999).

Synthetic Nitrogen Fertilizer

From Vaclav Smil we learn the role that synthetic fertilizer played in the extraordinary growth in population and even more incredible growth in food supply. In 1900, organic "virtually fertilizer-free" agriculture prevailed and was able to feed 1.6 billion people on a total of about 850 million hectares of land. "The same combination of agronomic practices extended to today's 1.5 billion hectares would feed 2.9 billion people, or about 3.2 billion when adding the food derived from grazing and fisheries" (Smil 2001, 159; see also D. Avery 2002). "Only about half of the population of the late 1990s could be fed at the generally inadequate per capita level of 1900 diets without nitrogen fertilizer. And if we were to provide the average 1995 per capita food supply with the 1900 level of agricultural productivity, we could feed about 2.4 billion people, or just 40% of today's total" (Smil 2001, 160).

Simply stated, "human activity has taken over from nature as the dominant source of fixed nitrogen in the environment" (Harrison and Pearce 2000, 75). Currently, "natural sources from soil bacteria, algae and lightning release 140 million tons of fixed nitrogen a year; human sources now total 210 million tons per year, of which 86 percent comes from agricultural activity, with fertilizer responsible for most of it" (Harrison and Pearce 2000, 75). Throughout the nineteenth and twentieth century, plant breeding was also making its contribution to increased production. Beginning during World War II with wheat research, followed by rice research beginning in 1960, the improved crops of what came to be called the green revolution provided the plants that could utilize synthetic fertilizer (and use it more efficiently) to produce ever greater outputs. The revolution in wheat and rice was preceded by the development of hybrid corn in the 1920s.

Norman Borlaug said in his Nobel Prize acceptance speech: "If the high-yielding dwarf wheat and rice varieties are the catalysts that have ignited the Green Revolution, then chemical fertilizer is the fuel that has powered its forward thrust. . . . The new varieties not only respond to much heavier dosages of fertilizer than the old ones but are also much more efficient in their use" (Smil 2001, 139). "The old tall-strawed varieties would produce only ten kilos of additional grains for each kilogram of nitrogen applied, while the new varieties can produce 20 to 25 kilograms or more of additional grain per kilogram of nitrogen applied" (Smil 2001, 139).

"Optimizing grain yields" in paddy rice production has the additional benefit of reducing the greenhouse gas methane (CH_4) emissions from the paddy fields (Denier et al. 2002). New hybrid varieties, of "rice, wheat and other crops" has the potential of increasing yields another 20 percent (Huang, Pray, and Rozelle 2002). New, more complicated breed hybrids are already in the fields in China with 20 percent gains in yield. Many NGOs now oppose the use of hybrids, calling them "an unfolding threat" to biodiversity (IPS 2002).

Many times I have heard the assertion that HYVs (high yielding varieties) require more fertilizer. This was often stated, as if the speaker was revealing the dirty little secret of the green revolution. The fact is that this statement has all the profundity of saying that it takes more food to raise three children than it takes to raise one since the HYVs required more nutrient because they produced a greater output. But for any given amount of nutrient input, the HYVs consistently produce a larger output than the conventional varieties because of their greater efficiency. Beyond a certain level of nutrient input, the conventional varieties cannot further respond without the large grain head lodging—bending over and no longer being able to mature. The HYV dwarfing varieties with their shorter, thicker stalks can continue to respond to greater fertilizer inputs beyond those of conventional varieties, producing an even greater yield.

Synthetic nitrogen fertilizer costs money, so as one would expect, farmers attempt to become more efficient in its use. The best measure of this is the ratio of nitrogen in the fertilizer applied to the nitrogen in the crop. This ratio fell for American farmers by 2 percent a year from 1986 to 1995. Further, there is no evidence that bulk deposition of nitrogen, which is of environmental concern because of runoff into rivers and streams, has been increasing (Frink, Waggoner, and Ausubel 1999). Another measure of increasing efficiency in nitrogen use is the feed-to-meat ration. As we have just shown, the synthetic nitrogen-to-nitrogen in the crop has been falling and now, in turn, the "calculated feed to produce a unit of meat fell at an annual rate of 0.9%" from 1967 to 1992 (Waggoner and Ausubel 2002). With increasing crop yields per acre, "cropland for grain-fed animals to produce meat for Americans *shrank* 2.2% annually" (Waggoner and Ausubel 2002).

Contrary to much popular misinformation, the major beneficiaries of the green revolution, this triumph of science and technology in the form of synthetic fertilizer and plant breeding have been the poorest and most vulnerable of the world's population. In the 1950s, the greatest concern about feeding people was in Asia whose history of famines stretched back for centuries if not longer. Rice is the staple of Asia and about half of the world's population eats some rice each day. From 1961 to 1991, "the population of Asia's developing nations more than doubled, from 1.6 billion to 3.4 billion. . . . However, per capita rice production grew by an impressive 170 percent, from 199 million tonnes in 1961 to 540 million tonnes in 2000, thanks largely to the introduction of improved rice varieties" (IRRI 2001). Globally, much the same story can be told. "There is currently an average of 2790 calories of food available each day for every human on the planet—23 percent more than in 1961 and enough to feed everyone" (IRRI 2001). Those who have fought and maligned the technology and science that have brought this about and who are the intellectual inheritors of those who over most of the last two centuries have opposed the development of this science and technology, now hypocritically use these data about its fruits to argue against the continued technological changes necessary to feed the nine billion expected by 2040.

With the real price of rice falling by 50 percent or more over the last four decades, a major beneficiary has been the very poorest consumers in Asia. This gain does not even take into account what the price and availability of rice would be if population had grown and rice production had not or at least not grown as rapidly. The subsistence farmer who rarely had little if any rice to sell, now has far more rice to feed his or her family and possibly some surplus to sell. The least beneficiaries have been the larger farmers whose gains in yields have been partially offset by the declining price for the sale of their output. Overall, "gains in food availability have been greatest in the developing world," where there has been a 38 percent increase in per capita calorie consumption "between 1961 and 1998 to 2660 calories per person daily" (Harrison and Pearce 2000, 59; Goklany 2002, 7, gives the figures of 2,257 in 1961 increasing to 2,808 in 1999). From 1964–66 to 1997–1999, 57 percent of the "world's population were living in countries with average intakes of 2200 kcal per day," a proportion that has declined to about 10 percent while the absolute number "fell by two-thirds, from 1,890 million to 570 million" (FAO 2002, 14). Developing-country consumers are also catching up in meat consumption as their demand for meat and poultry is growing at "twice the rate of population growth" (Harrison and Pearce 2000, 59).

A belief popularized by vegetarians and those opposed to modern agronomy is that the world could better feed everyone on less land under cultivation if we all gave up eating meat or at least greatly reduced our consumption of it. As I have argued previously, humans are inherently meat and fruit eaters as our large brains require an amount of energy that our digestive system with relatively small hind gut could not process except for the provision of food that is more energy dense than grains and vegetables (DeGregori 2001, 77–78). Thanks to modern food processing and chemistry, which the vegans scorn, they can obtain more calorically and nutritionally dense foods than can be found in nature and to obtain essential nutrients such as vitamin B_{12}, which previously could only be obtained from animal products.

Animals "eat huge amounts of forage that humans cannot digest, from grasslands that mostly cannot support crops" and "such high-yield forages as alfalfa, which produce much more biomass per acre than the food crops that might replace it" (D. Avery 2002). Coarse grains used for animal feed such as maize and sorghum achieve higher yields because of more efficient photosynthesis as a result of having the genes encoding the C4 enzyme instead of the gene for the C3 enzyme characteristic of most other grains and vegetables (Ronald and Leung 2002). "In addition, animals and poultry eat millions of tons of such by-products as distillers' dried grains and millers' wastes which humans can't digest" (D. Avery 2002). In citing pounds of feed to pounds of meat, vegetarian advocates neglect to consider other issues of the caloric nutritional density, digestibility and overall quality of the inputs compared to the quality of output. A report for the UN Food and Agriculture Organization found that "animals worldwide consumed 74 million tons of human-edible protein and produced 54 million tons of human food protein. This gives an input:output ratio of 1.4 to 1. As it happens the ratio of biological value in animal protein compared to plant protein is also 1.4 to 1" (D. Avery 2002).

Being a vegetarian may also be harmful to biodiversity and be fatal to more animals than eating meat. "The grain that the vegan eats is harvested with a combine that shreds field mice, while the farmer's tractor crushes woodchucks in their burrows." If we were to "adopt a strictly vegetarian diet, the total number of animals killed every year would actually increase, as animal pasture gave way to row crops. . . . If our goal is to kill as few animals as possible, then people should eat the largest possible animal that can live on the least intensively cultivated land: grass-fed beef for everybody" (Pollan 2002).

The living molecule "represents the coming together of atoms in proportions that are, if not constant, at least bounded and obeying some rules." They obtain the atoms "from an environment that may or may not have similar proportions of these elements" and the molecular compounds that they form with other elements (Sterner and Elser

2002, 3–7). In many respects, this is Liebig's Law of Minimum, which says that "organisms will become limited by whatever resource is in lowest supply compared to their needs" (Sterner and Elser 2002, 81).

The land that humans first cultivated rarely had the necessary proportions of the elements necessary to feed their crops so that humans had to devise means of amending the soil. As agriculture developed on into modern times, the more crop that was harvested meant the more elements that are the constituents of plant and human nutrient were being taken away and had to be replaced. Feeding six billion today and nine billion in 2050 and feeding them better than ever before means more nutrient taken from the land and therefore more that must be returned.

Modern synthetic fertilizer may be largely nitrogen based but any modern regimen of fertilization seeks to replace the various nutrients that previous cropping has removed or may have not been available in the first instance. An organism "must take complex chemical resources containing multiple elements arranged in myriad different molecules, absorb some, metabolize some, rearrange many, and excrete or otherwise release a great deal" (Sterner and Elser 2002, 3-7). There are still many places where human and animal excretion is returned to the land but given that even in the best of circumstances, recycling is never 100 percent, synthetic fertilizers will be playing an increasing role in feeding the world's population.

Allergenic Proteins and Amino Acids

Twenty common amino acids are used to create proteins for all life forms on earth. Though some proteins nourish us, other proteins can kill us. Proteins that nourish some humans—those in peanuts, for example—can cause a fatal allergenic reaction in others called anaphylactic shock. Quite literally one person's meat (or nutrient) is another person's poison. Ninety percent of food allergies in Western countries result from eight food types: peanuts, wheat, soybeans, milk, tree nuts, eggs, shellfish, and fish. The common element in these foods is a relatively indigestible protein.

The proteins most likely to be allergenic are those that come from a food that itself causes allergies, has similarities in its sequencing to known allergenic proteins or does not dissolve in stomach acid at a rate that would break it down into its amino acid components before being passed on to the rest of the GI (gastro-intestinal) track. Many food allergies result from bits of undigested proteins that continue in the digestive track and remain there for enough time to induce the production of IgE antibodies. IgE antibodies produced on the first encounter will recognize the same protein on repeated encounters with a greater response each time, which can trigger an allergenic reaction that can vary from being mild to severe to fatal (Sicherer 2002). "Proteins that resist digestion often do so because they are folded too tightly for digestive enzymes to gain access" such as in a protein with a disulfide bond (Day 2001). Some of the same plants that have the proteins with disulfide bonds or other tightly folded bonds also produce thioredoxin, "a non-allergenic protein that specifically targets chemical bounds that hold proteins in their folded shape" (Day 2001). One of the aims of agricultural biotechnologists is to increase the thioredoxin production of a plant such as wheat and therefore make it less allergenic (Day 2001). The first harvest of a hypoallergenic wheat was to take place in 2001 and be tested for effectiveness (Day 2001). For some with allergies like many have for peanut protein, increasing the thioredoxins won't make it safe to eat but might offer some relief from the continuing threat of a possible fatal reaction from even the slightest accidental encounter or the minutest contamination. "Gene silencing" can silence genes that express an allergenic protein. This has already been done for the genes for the allergenic P34 protein in soybeans responsible for half of soy allergies (USFDA 2002).

Bioengineered Bt (*Bacillus thuringiensis*) corn has a protein that is activated by enzymes in the insect gut when ingested by the corn borer or other insect pests. The activated Bt protein binds to specific receptor sites in the gut and inserts itself into the membrane of the insect gut. Bound to the inner linings of the stomach, the Bt toxin causes a influx of water into cells that swell and destroy the insect digestive system (Nester et al. 2002). "As the gut liquid diffuses between the cell, paralysis occurs, and bacterial invasion follows" (Benarde 2002, 117). This leads to insect starvation and eventual mortality and is the same mechanism used by the live *Bacillus thuringiensis* bacteria to kill the insect and then feed and multiply on its remains. The Bt protein does no harm to "birds, fish, or mammals, including people" (Benarde 2002, 117). The stomach of vertebrates including humans is acidic; those of insects (arthropods) are alkaline. "Since Bt's crystalline protein is alkaline, it can function effectively at the higher pH range. Receptor sites for this protein are lacking in acid environments; therefore, Bt is harmless to all but insects" (Benarde 2002, 117).

We refer to a "specific receptor" for the Bt toxic protein. In the interaction between plants, insects, and microorganisms, and the adaptations that are made between predators and prey or between plants or animals and parasites, some insects develop adaptations in which they feed on a specific plant or type of plant while some microorganisms specialize on infecting specific organisms for their survival. Specialization in nature, like other forms of specialization, limits the options of the organism but gives it an advantage in exploiting the one to which it has adapted. A plant or insect subject to attack by a specific insect or parasite will tend to develop resistance to it. In the struggle for survival in nature, the emergence of a trait that improves the ability to resist predation or to prey on others will spread through the species, becoming dominant.

It is a truism that the plants that develop the trait or complex of traits or defense mechanisms that most effectively allow them to ward off predators are the ones that survive. Plants, both natural and domesticated, are chemical factories that produce the vast majority (over 99.9 percent) of the toxins that we ingest each day. Periodically we are offered non-peer-reviewed studies that claim to prove the nutritional superiority of organically grown produce on the basis of a greater production of many of these chemicals, which they label "phytonutrients" and which a plant physiologist would call secondary metabolites (since they are not essential for basic plant metabolism). It is unlikely that the plant's genes would be instructed to express these chemicals out of gratitude to the farmer for nourishing them with manure instead of synthetic fertilizer since the manure has to be broken down into compounds (cations—an ion with a positive charge, anions with a negative charge, and uncharged particles) that are no different from those in synthetic fertilizer. A more likely explanation is that the plant was engaging in this otherwise wasteful production of manufacturing toxins to protect against fungal or other infestation (Mattisson 2000). The possibility is not considered by these so-called studies that the chemical compounds might not be nutrients but might be harmful to humans or even be carcinogenic (Ames, Profet, and Gold 1990a,b).

Reports by scientific organizations arguing no difference between organic and conventional food (except for the organic's "potential for greater pathogen contamination") are ignored as are those that find a potential for greater nutrition in transgenic crops (IFST 2002, Demmig-Adams and Adams 2002). In August 2001, a study purporting to offer scientific evidence for the superiority of organic produce was announced by the Soil Association in the United Kingdom and was parroted by the media, which repeated the claim of scientific evidence where there was none. No research was involved in a survey of over four hundred existing reports carried out by someone who was not a scientist, though it was implied that he was. His training included a bachelor's degree in business and study at a cult science institute staffed by its own graduates. The study purported to find evidence for the superiority of organic food, which the original researchers failed to find. The original four hundred were first reduced to ninety-nine of which only twenty-nine were deemed "valid" or not "flawed" including ten by organic food enthusiasts, which were not peer-reviewed (Avery and Avery 2002). Being flawed meant a failure to find any evidence for the superiority of organic food. Some of the studies were termed "downright hocus-pocus." "So-called 'holistic' methods of food quality analysis look at the crystal patterns formed by copper salts in the juices of fruits and vegetables. Organic activists claim these tests show organic foods have better 'picture forming' qualities, and thus more 'vital quality'" (Avery and Avery 2002).

Avery and Avery ask us to excuse their "snickering" for which they in fact have good reason because more than seventy years after "the first copper crystallization tests were conducted, not a single food scientist can say what the crystal formations mean in terms of food quality or nutrition. It is quite literally pseudo-scientific mumbo jumbo, but it surely helps promote organics to new-age consumers" (Avery and Avery 2002).

The valid twenty-nine include a "research paper by Dr. William Lockeretz of the Tufts University School of Nutrition Science . . . a long-time organic proponent and a co-founder of the pro-organic *American Journal of Alternative Agriculture.*" Avery and Avery cite a 1997 statement by him "at an international organic conference": "I wish I could tell you that there is a clear, consistent nutritional difference between organic and conventional foods. Even better, I wish I could tell you that the difference is in favor of organic. Unfortunately, though, from my reading of the scientific literature, I do not believe such a claim can be responsibly made" (Avery and Avery 2002).

The organic enthusiasts never seem to tire of trying to find evidence of the superiority of their product. In March 2002, another study was announced that purported to show that organic vegetables were more nutritious than those that were conventionally grown. This study even appeared in a peer-reviewed journal, which is a rare occurrence for such studies, given the paucity of evidence (Baxter et al. 2001). Canned soups made with organic vegetables were found to have a higher level of salicylic acid than vegetable soups that were not labeled organic. Salicylic acid is the active ingredient in aspirin, which has been shown to have beneficial health effects for those who take one or two a day. This was taken as tantalizing evidence of nutritional superiority, which warranted further research. No matter how heroic the efforts to hold other factors constant, it is difficult to take seriously a study comparing commercially canned vegetable soups where one could either have compared the vegetables directly or made the soups themselves to make sure that everything else was the same except the vegetables.

Even if one accepts the study's validity, its conclusion is still in doubt. The tiny amount of salicylic acid that they found was 117 nanograms/gram. This is "1/10,000,000 of a gram or 0.00001% or (1/100,000 of 1%). For a typical 400 gram serving of soup at 117 nanograms/gram = 50,000 nanograms of SA, which is 0.005% of a gram, or 0.05 milligrams of salicylic acid, or 1/20th of one milligram" (Avery 2002a). A bowl of organic soup provides "roughly 1/6,000 of a standard aspirin compared to conventional soup," which provides "only 1/36,000 of an aspirin" (Avery 2002a; see also Fienberg 2002). A grandiose conclusion from virtually no evidence is an example of what Irving Langmuir called "pathological science" (Langmuir and Hall 1989).

Despite the small amounts of salicylic acid, there was a posted warning (at AgBioView and subsequently at AgBioWorld.org) about the dangers of eating organic food, citing testing by the U.S. Food and Drug Administration and the American Medical Association, as excess consumption may cause "bleeding problems or gastric ulcers and should be avoided by pregnant women, nursing women and children under 12." The posting gave warnings to many groups that they should avoid eating organic vegetables. I thought the posting was a spoof, turning the logic that even barest trace amounts of a chemical are dangerous, used by the organic proponents against them until I received an email from a research scientist in Australia "with an acute sensitivity to salicylates. These minute amounts can mean the difference for me between having a serving of vegetables every day or only once a week." She adds: My dietician told me not to "eat organic foods precisely because they contain higher amounts of salicylates. She told me this fourteen months ago, so it has been common knowledge among dieticians. She said the three most common causes of food sensitivities (not allergies) are salicylates, amines, and glutamates, which are all found naturally in foods. They are found in higher levels in organic foods because they are produced to protect the plants" (Brumbley 2002).

The last sentence is revealing since the article admitted that the lower level of protection against microorganisms and disease bearing insects in organic agricultural plants causes them to produce higher levels of secondary metabolites such as the salicylates. Since most secondary metabolites are rodent carcinogens, maybe the same researchers ought to have tested for other secondary metabolites though it is unlikely that such tests would get the same media attention if they found carcinogens. Since the plants produced them, it is reasonable to conclude that they experienced greater infestation by fungi or other harmful organisms. It would be fair to conclude that the study is trivial for those who need salicylates and, contrary to the claims, it is dangerous for those with acute sensitivity to salicylates. In other words, the study demonstrates the exact opposite of what it claims and what is being claimed for it (DeGregori 2002a,c).

If customers want aspirin in their soup, it would be quite cheap for producers to add all they wanted and still sell their soup for less than the organic brand. The manufacture of vitamins or common pharmaceuticals such as aspirin is incredibly cheap (particularly when made and sold in bulk to producers) and they can then be added to food. For nutrition intervention programs (such as vitamin A) in poor countries, the cost is not in the vitamins themselves but in creating a delivery mechanism. Of course, these nutrients lack the mystic potency and life force of the high-priced, similarly manufactured vitamins in a bottle labeled "all-natural." The next project of our intrepid researchers ought to compare the nutrients in Total (a cereal with 100 percent of the RDA of all nutrients thanks to modern chemistry) to an organic cereal.

The human body takes in a variety of complex proteins that it then breaks down into amino acids, which it uses to build its own protein. As long as it has a sufficient number of the essential amino acids, the body can synthesize the other amino acids necessary to create the proteins that constitute us. Given that there are twenty amino acids common to all life including our own, as long as the protein that we ingest (be it natural to the plant or bioengineered into it) is fully broken down in the stomach, it can not be allergenic or otherwise harmful. And until the proponents of organic agriculture can demonstrate that their plants produce a greater amount of the nutrients required for humans without producing more of that which is harmful to us, we will have to judge their claim of a nutritionally superior product to be not proved.

Agriculture, Crop Protection, and Food Safety

To the extent that herbicides are not used in organic agriculture, eliminating the weeds requires stoop labor in the hot sun, often by exploited migrants who "toil in dangerous, unsanitary conditions" often for wages less than the "legal minimum" (Roane 2002). Many who buy organics assume "that growers treat their workers with as much care as they do their tender shoots and berries" (Roane 2002). This perception is "fed by an $8 billion-a-year industry that increasingly touts the organic label as a lifestyle that transcends mere food" (Roane 2002). "Organic crops must be weeded far more often than crops treated with chemicals, and after California banned short-handled hoes as dangerous to workers' backs, some organic farmers sent laborers out with no tools at all, forcing them to hunch over for hours in the baking sun. Long hoes would allow workers to stand upright, but some farmers believe these tools can damage crops" (Roane 2002).

The much celebrated USDA rules for organic food labeling provide a substantial list of approved pesticides both natural and synthetic (USDA 2002). The rules for organic chickens require them to be "free range," which exposes "the birds to rats and mice that carry salmonella bacteria and to ducks and geese, which can spread bird diseases such as avian influenza." One believer argues this is "necessary to ensure chickens live as normal a life as possible." He adds: "What's natural for a chicken is the best practice" (Brasher 2002).

Even protected plants produce some chemical defenses though fewer than the same plant if unprotected. Those plants that have survived in nature have done so because of the successful chemical and other defenses they have evolved. Domesticated plants that have long been removed from the habitat of their origin and the predators therein, often lose the ability to produce the specific chemical and other defenses since they would not have any survival value and would even be energetically wasteful. This explains why farmers and plant breeders seek plants from the original habitat for crossbreeding for resistance when a new disease or predator invades their domain. Given that this process of developing resistance and new forms of attack followed by new resistance and new means of attack in a seemingly never-ending process, it is understandable that with human intervention in the form of domestication, there is the same process of chemical or biological defense against insects and microorganisms, followed by the evolution of means of overcoming these defenses in an ongoing process. Critics of modern agronomy, in recognizing this process, offer a perverse form of Luddite logic in concluding that no defense should have ever been tried since the insects or microorganisms would eventually evolve means of overcoming them. How we would be better off by never having tried to protect the crop is never fully explained.

Contrary to the doomsayers, some of the modern commercial plant varieties that have had resistance genes bred into them have maintained this resistance for long periods of time—up to fifty years in some cases—and are still functioning well. "In the United States, the T gene in barley has held up against stem rust for over 50 years; similarly, in wheat the Hope gene has kept stem rust in check for over 40 years and the LR-34 gene has limited leaf rust for more than 20 years" (Sanders 2000).

To Sanders, "multiple-gene resistance and other techniques are preferable when they are available" but we "use what we have if it works, and we anticipate breakdowns" (Sanders 2000). This pragmatic process of breeding in plant protection is not only vital for agriculture; there is no alternative to using a variety of modern crop protection strategies.

When two research reports and a news of the week article on the development of resistance to the Bt toxin were posted online in *Science* for the August 3, 2001, issue, antibiotechnology groups almost instantly picked on the recognition of the Bt resistance and were online with it in their campaign against genetically modified food before most subscribers even had the hard copy in hand. The online postings were quickly followed by news stories strikingly similar to the anti-GM postings. A close examination of the articles (or even a cursory one) would have indicated that an understanding of them would not advance the cause of those against the use of biotechnology in agriculture.

First, the Bt "resistant strains of at least 11 insect species have been documented in the laboratory" while only "Bt-resistant variants of the diamondback moth have been identified in the field" (Griffitts et al. 2001). Checking the article footnoted for resistant strains found in the field indicates that they occurred before 1994, the date of the cited article, which was also before the first Bt modified varieties were released (Griffitts et al. 2001). In fact, resistance to live Bt spray by the diamondback moths emerged in the field as early as 1989 (Palumbi 2001). "Some populations of diamondback moths, a devastating pest of cabbage and related crops, are no longer bothered by sprays of Bt bacteria used by organic farmers" (Stokstad 2001). In other words, the use of the live bacillus has the same potential of creating resistant strains as does the use of the toxin engineered into the plant, though obviously the more extensive use of the Bt toxin in any form will likely accelerate the development of this resistance. But note again, the only resistant strains mentioned in the articles that were found in the fields were found in those involving organic agriculture.

Those in the environmental movement who oppose the patenting of life forms somehow believe that organic farmers have an exclusive absolute property right to use and prevent others from using not only the live bacillus but also the protein toxin that it produces because of their hallowed tradition of fifty years of use. The three articles in *Science* reveal a critical difference between those who favor the use of science in agriculture and those who would favor some other method. Modern agronomy provides a variety of strategies for agriculturalists to employ, in addition to Bt, such as chemical pesticides and refuges to maintain a population of insects that do not develop a resistance to the Bt toxin. The articles demonstrate that modern biotechnology provides the ability to identify and monitor "resistance allele frequencies in field populations," so that farmers will have a "direct test of whether the high-dose/refuge strategy is succeeding." This "may allow enough time for the strategy to be adjusted to reverse the increase" if the existing strategy "starts to fail" (Gahan, Gould, and Heckel 2001; see also Ferre and Van Rie 2002). The articles indicated that insects were evolving defensive mechanisms, which presented a challenge to create new strategies to combat them.

Those who read the online environmentalist postings would have never surmised that the authors of one of the articles were defining ways of facilitating the long-term use and expected benefits of Bt-engineered crops. This is clear in the following concluding reference on "the opportunity to make informed modifications to a strategy that could sustain the use of Bt transgenics and prolong their environmental benefits of reducing dependency on conventional insecticides" (Gahan et al. 2001).

Those who oppose all uses of biotechnology in agriculture, deeming it to be inherently evil, lack any realistic options to counter the growth of resistance to live Bt spray. Biotechnology and agronomy, like all scientific inquiry, are processes of inquiry (the scientific method) and problem solving. They are in search of the best solutions to problems, not the ultimate solutions. In some cases, such as that of live Bt spray and the T gene in Barley, the solution works for a long time. In others, the time frame is much shorter. The critical difference between science and the presumed alternatives is that science has a way of moving forward to find solutions and even to anticipate a need for them (Mokyr 2002, 38). Considering the way in which opponents of Bt corn have been characterizing its threat to organic farmers using live Bt spray, one might surmise that the organic farmers could continue using it in perpetuity were it not for the intrusion of the bioengineered Bt serpent into their Edenic preserve.

Bt Corn and the Monarch Butterfly

A much publicized correspondence to *Nature* suggested that "transgenic pollen harms Monarch larvae" (Losey, Rayor, and Carter 1999). The study was carried out in the laboratory with monarch butterfly larvae placed on a leaf that had been dusted with pollen from Bt corn. The correspondence had been rejected as an article by the two leading science publications and was contradicted by previous field studies (Shelton and Sears 2001; Shelton and Roush 1999). The Luddites immediately were able to generate massive and continuing publicity for this rather insubstantial evidence of harm. Contrary to much publicity and street theater, the monarch butterfly is unharmed by ingesting the Bt protein at the levels to which it is exposed, as has been recognized by the EPA, even though the monarch butterfly does have the appropriate receptor for the Bt toxin. That butterfly larvae could be harmed by the Bt protein if ingested in sufficient quantities, should not have come as a surprise; the live Bt is used in organic agriculture to protect the crop against other insects of the same order as the monarch butterfly, Lepidoptera. Several major peer-reviewed studies were published in PNAS (Proceedings of the National Academy of Sciences of the United States of America) that support the EPA conclusion that the Bt protein is safe for the monarch and its larvae at the level of exposure to which they are subjected by the Bt corn products currently in the market (Hellmich et al. 2001; Pleasants et al. 2001; Sears et al. 2001; Stanley-Horn et al. 2001; Zangerl et al. 2001). The evidence continues to grow with yet another study in a peer-reviewed journal finding that Bt maize hybrids did not "pose a significant risk to the monarch population" and Bt-expressing crops posed "little risk to other nontarget insects, including beneficial insects such as pollinators and natural enemies" (Gatehouse, Ferry, and Raemaekers 2002). Some strains of live Bt produce additional "much more broadly toxic" exotoxins and enterotoxins than the endotoxin of transgenic maize (STPP 2002). A 40 percent plus in Bt maize acreage was accompanied by a 30 percent increase in monarch according to the highly respected Monarch Watch website—http://www.MonarchWatch.com. In fact, "GM technology has the potential to contribute to the preservation of biodiversity relative to other management practices" including organic (Gatehouse et al. 2002).

As with most antitechnology mythology, massive refutation may dampen the furor it caused, but substantial evidence to the contrary rarely extinguishes the myth entirely as it lives on in the nether world of the believers. With the PNAS reports, the environmental case against Bt corn clearly collapsed. Rather than accept the obvious fact, the anti-GM food groups asked the U.S. EPA (Environmental Protection Agency) for more time to make their case against renewing the license to plant Bt corn. The critics keep asking for more proof for the safety of GM crops and food but the more you give them, the more proof they demand. There is simply no amount of proof that they will ever find satisfactory or compelling. Evidence becomes irrelevant when evil is deemed to be inherent and eternal. The call for more evidence is simply a ruse to make the critics appear open to evidence when they are not. When the UNDP (United Nations Development Programme) in its *Human Development Report 2001* recognized the potential benefits for poor countries of genetically modified crops in a balanced presentation, the critics responded by vilifying the report rather than engaging in a reasoned dialogue with an organization whose concern for the very poorest and most vulnerable humans on this planet is essentially beyond reproach (UNDP 2001).

The preponderance of evidence from field studies and from the record numbers of monarch butterflies indicates that, on balance, Bt corn is less harmful to butterflies than alternative kinds of corn. Further, with new varieties of Bt corn that have the protein toxin primarily in the stalks and leaves with very low levels in the pollen, the advantage of the Bt corn over other kinds of corn becomes even greater.

Agriculture and Nature

The word *natural* has been overused and misused so much as to have become virtually meaningless. If the term is to have any meaning, then what could be more natural than insects becoming resistant to methods of controlling them? Modern agronomy and biotechnology do not seek to repeal the principles of evolution but to make use of them. Agriculture is too important to all of our lives, in addition to its being the livelihood of those who grow our food, to allow a variety of insects or microorganisms or other threats to crops, to destroy field after field until a few hardy mutants remain, allowing cultivation of a crop to continue. Using that energetically expensive organ, our brain, we have become a vital part of the evolving interaction between plants and predators in the crops we cultivate. Domestication has made humans an integral part of the evolutionary process of resistance and overcoming resistance. This is far more natural in line with our understanding of biological evolution than simply trying to defend one's turf of a strain of Bt spray against what in the long run will likely be the evolution of resistance to it, an event that has already begun to happen.

With the exception of the Starlink protein for which the data are incomplete, the protein in Bt corn is fully digestible by humans. As such it is broken down in the stomach into its amino acid components, thus the importance of the twenty (or slightly more) amino acids common to all life. While the body may recognize a protein as being alien and have an adverse allergenic reaction, there is no such allergenic response if the protein is broken down to its constituent amino acid components before entering the intestines. With all the fuss over the possible allergens in transgenic food, it should be noted that for almost all known allergenic foods, the body builds up more serious allergenic responses to it after repeated exposure to it, with some becoming potentially fatal. Rarely is the allergenic response serious or fatal at the first encounter. One rare exception is the allergenic response to the herbal remedy echinacea, beloved by the devotees of alternative medicine and natural remedies. There have been severe allergenic reactions to echinacea on what appears to be the first exposure to it by a number of subjects. Echinacea is a member of the ragweed family. The data support the "possibility that cross-reactivity between echinacea and other environmental allergens may trigger allergenic reactions in 'echinacea-naive' subjects" (Mullins and Heddle 2002).

Referring to the protein in the Bt corn as being a toxin or a pesticide is technically correct as it is toxic to the target species, the corn borer, which it kills, making it therefore a pesticide. But it is misleading to characterize the protein as a toxin and pesticide when discussing the Bt corn as a human foodstuff, if it in fact completely breaks down in the stomach and can therefore cause no harm. (In an activist posting, I saw the Bt protein repeatedly referred to as a toxin in discussing Bt corn and then later in the same article, saw it called a protein in reference to live Bt.) With the bioengineering of foodstuffs offering such great possibilities for human betterment, and with a protein often being the bioengineered component, it is clearly necessary to take care to avoid making a previously safe product allergenic (Palevitz 2002b). This is done for all bioengineered foods, which have to be tested in various ways for safety before being released. It is equally important that the product not be misrepresented by those with a vested interest in opposing it.

Another Controversy

Two and one half years after the Losey correspondence on Bt corn (see above), a letter to *Nature* raised another furor by Non-Governmental Organizations opposed to transgenics in agriculture (Quist and Chapela 2001). Quist and Chapela claim to have found evidence of transgenes in native Mexican maize landraces and claimed that they had jumped about in the genome and inserted themselves in a "diversity of genomic contexts." Immediately cries of "contamination" and "genetic pollution" arose along with calls for a global ban on the planting of Bt corn. Unfortunately, the media tend to use the loaded and distorting terminology of the critics of modern technology, giving undeserved credence to their claims (Felsot 2002). Given the method used, iPCR (inverse Polymerase Chain Reaction) is subject to false positives, scientists were skeptical of the findings of contamination but conceded that if it has not yet happened, it will eventually do so. It was only later revealed that Quist and Chapela sent samples to a scientist in the Swiss Ministry of Health who tested them and found the results to be so "faint" that they could not be considered positive and therefore it was not publishable (Ammann 2002).

It is amazing how those opposed to modern science and its methods are not only willing to use research techniques such as PCR, iPCR, and ELISA (Enzyme-Linked Immunosorbent Assay) but are also willing to make claims of certainty far beyond what those who developed and use them regularly in their scientific work would claim. Apparently scientific research methods are to be distrusted and can be totally discounted and ignored unless they further an antiscience ideological agenda. Any questioning of adverse findings is immediately labeled a smear.

As with the Losey episode, there was already quality peer-reviewed research showing that even with the inevitable movement of transgene into the native landraces or teosinte, there was "no need for concern" (Martinez-Soriano and Leal-Klevezas 2001). Those engaged in plant research were even more skeptical of the claim that the transgene moved about and inserted itself in a "diversity of genomic contexts" since little evidence was offered and it was contrary to the research experience of those involved in plant genetics (Christou 2002; Conko and Prakash 2002; Hodgson 2002a,b; Felsot 2002; and Martinez-Soriano et al. 2002). Further, Quist and Chapela refused to release data on their control other than to insist that they were negative. They were also accused of misciting at least one reference (Salleh 2002).

Several scientists sought to obtain samples of the material in which the so-called contaminants were presumably found, requests that were denied. A hallmark of modern science is its public character and its replicability (Bronowski 1965). The two go together, as a scientific article is required to contain enough information as to how it arrived at its conclusions so that others can replicate the procedure and achieve comparable results. For some research, provision of the research materials is required for replicability to be possible, and reputable journals have requirements to assure that this happens. "When authors submit a manuscript, they make a commitment to supply cells, special reagents, or other materials necessary for verification" (Kennedy 2002). "They are not free to violate that commitment once their paper has been published. *Science*'s Instructions to Contributors set out the rule this way: 'Any reasonable request for materials and methods necessary to verify the conclusions of the experiments reported must be honored'" (Kennedy 2002).

When a respected organization like CIMMYT (Spanish acronym for International Center for the Improvement of Maize and Wheat) analyzed its accessions and could not verify the claims, it and CGIAR (Consultative Group on International Agricultural Research) were essentially accused of cowardice by being silent on this issue. CIMMYT continued to search for evidence of contamination by transgenic maize and repeated its claim that it found no evidence of an intrusion (CIMMYT 2002; CBU 2002).

It was a Nobel Prize winner, Norman Borlaug, working at the predecessor to CIMMYT (which he was instrumental in founding) that created the HYV (high yielding variety) wheat, and it was CGIAR members such as IRRI (International Rice Research Institute) that created the other HYVs (or disease-resistant varieties) such as those in rice, which have allowed the earth's six billion people to feed themselves far better than the three billion were able to do just four decades ago. When you see, on the evening news, large planes off loading sacks of wheat in response to some disaster that results in famine, think of Norman Borlaug, CIMMYT, modern agronomy, and the high-tech conventional mechanized agriculture that produced the surpluses. When you see pictures of famine in Asia in times past and see today a population that has more than doubled in the last half century and is better fed than ever, think of IRRI and the high-yielding varieties of rice and the green revolution. When the critics of CIMMYT and IRRI purport to take a moral high ground on issues of food and agriculture, one has the right to ask them how many people, if any, in this world are better fed because of their efforts?

Those who raised honest scientific arguments against Quist and Chapela's claims were accused of slander and intimidation though no evidence was offered for anything but objections to the research and publication based upon issues of proven scientific methodology (Palevitz 2002a). The purity of the environmentalist cause makes anyone supporting it beyond reproach and any argument beyond challenge. Any counterargument is by definition, a smear.

Over and over in the press releases and postings, it was stressed that the Quist and Chapela article was peer-reviewed. Given the almost total absence of peer-reviewed articles in quality journals supporting their cause, some of us thought that the antitransgenic NGOs didn't believe in the process. When some of the most respected scientists doing research in this area raised a number of questions challenging the validity of the results and their significance, their integrity was questioned. They were accused of trying to suppress dissent and of carrying on a scurrilous smear campaign against a couple of honest researchers doing their duty defending Mother Earth's biodiversity. Among the so-called smears was the charge that one of the authors was a member of the board of an antipesticide organization and therefore not the disinterested researcher that he claimed to be. Had any of the critics even the remotest connection to a multinational corporation, such as receiving a research grant several years back, this would have been trumpeted loudly with a correlative charge of corruption and dishonesty by those groups claiming a smear against Quist and Chapela.

Contrary to this fiction, the criticisms of Quist and Chapela were scientific and not personal. Given that the ongoing antitransgenic NGO rhetoric means to demonize scientists who differ with them, inevitably there are claims of a corrupting corporate tie with implications of venality and dishonesty. Some note was made of the activism of both Quist and Chapela in opposition to transgenics. The activism of the authors was a minor element in the discourse and was not even alluded to by the scientists criticizing their correspondence. One searched in vain on the antitransgenic NGO websites and newsgroups for any attempt to respond to the scientific issues being raised.

Correspondence questioning the scientific validity of the Quist and Chapela piece poured into *Nature*. They were peer-reviewed and two were published (Metz and Futterer 2002; Kaplinsky et al. 2002). Quist and Chapela were allowed to defend themselves with additional research evidence (Quist and Chapela 2002). Their submission was also peer-reviewed and the scientific validity of the new evidence was rejected but was still published as was appropriate. Following each of the three letters was the same editorial note that made reference to the additional data and the referee's report and contained the following sentence: "In light of these discussions and the diverse advice received . . . the evidence available is not sufficient to justify the publication of the original paper" (Metz and Futterer 2002, 601; Kaplinsky et al. 2002, 602; Quist and Chapela 2002, 602). This is believed to be the first time in the 143-year history of *Nature* that it has questioned its original decision to publish a paper. Two letters were later published in *Nature* in defense of Quist and Chapela. They made the usual attacks against the scientists who differed with Quist and Chapela claiming that they were essentially servants of financial interests. They were also critical of *Nature*'s actions in suggesting that they were in error in originally publishing the Quist and Chapela piece. What they did not have is even the merest hint in defense of the scientific validity of Quist and Chapela's work (Nature 2002).

NGO enthusiasm for so-called peer-review and prestigious scientific journals suddenly evaporated but rhetoric about contamination continued undiminished. Scientists conceded from the beginning that some exchange of genes was inevitable but have been equally strong that this is quite natural and unlikely to cause any harm. The media continue to find it expedient to use pejorative NGO terminology, such as "contamination," when there are more accurate terms that would be more in line with good journalistic practice.

Like it or not, the issue has become one of science versus antiscience as one increasingly finds, among the antitransgenic NGOs and their brethren, blanket condemnations of modern science as being corrupt, not to be trusted, and beholden to corporate interests. Some of the posted antitransgenic rhetoric seeks to brand all of modern science as subservient to corporate interests except for that minuscule minority that supports their cause. This fits in nicely with the antiscience, antitechnology musings of the postmodernists and ecofeminists, which I have discussed.

Science as Ongoing Inquiry

Even if one peer-reviewed correspondence on contamination hadn't been so massively refuted, it would not have been as definitive as the anti-GMO (genetically modified organisms) campaigners implied. One peer-reviewed article or correspondence in the most prestigious journal is not as significant as it is often reported by the media. One author refers to this as the "creation myth . . . in which a few discoverers have great ideas and the scientific community immediately falls into their enthusiastic trial and acceptance" (Morgan 2002, 487–88). Morgan adds that "ideas are rarely accepted on the strength of a single paper, no matter how good it is. Only when enough trailblazers have mined a new idea does it begin to be accepted by the broader community" (Morgan 2002, 487–88). Science is a "construct of arguments and counterarguments that we try to fit together in a mental crossword puzzle" (Vandenbroucke and de Craen 2001, 511). As in a crossword puzzle, one judges new evidence in terms of already existing evidence in much the same manner that the new word is judged in terms of "already-completed entries" and "how well evidence supports a proposition depends on how much the addition of the proposition in question improves its explanatory integration" (Haack 1998). When a new theory is "still developing," scientists may remain committed to an existing theory if it is part of a larger body of knowledge that works and has worked "from its ancient forms in the construction of temples or cathedrals to telecommunications and gene therapy" (Vandenbroucke and de Craen 2001, 512).

A scientist has to be driven in part by the desire for truth and a belief that expanding truth or knowledge is beneficial to humankind. "Most daily activity in science can only be described as tedious and boring, not to mention expensive and frustrating." Few would do it without a recognized, shared purpose. In science there is a "core assumption about the existence of an accessible 'real world' that we have actually managed to understand with increasing efficacy, thus validating the claim that science, in some meaningful sense, 'progresses'" (Gould 2000). "How could scientists ever muster the energy and stamina to clean cages, run gels, calibrate instruments, and replicate experiments, if they did not believe that such exacting, mindless, and repetitious activities can reveal truthful information about a real world? If all science arises as pure social construction, one might as well reside in an armchair and think great thoughts" (Gould 2000).

Even if one accepts with some modification, the postmodernist claim of science being a social construct, this does not invalidate the Pierce/James/Dewey thesis that if an idea works, it is a validating argument for its truth or that its truth and its working out are one and the same thing. "The true, insightful, and fundamental statement that science, as a quintessentially human activity, must reflect a surrounding social context does not imply either that no accessible external reality exists, or that science, as a socially embedded and constructed institution, cannot achieve progressively more adequate understanding of nature's facts and mechanisms" (Gould 2000).

However boring the nitty-gritty of science may be, putting the details together as part of a larger social process makes us all partners in this most magnificent of human endeavors. It is in the social process of science that knowing and doing, truth, knowledge, and problem solving become integrally intertwined. Together we can "build a maximally reliable vehicle for this most adventurous of all improbable journeys toward the grandest goal of human striving and natural order" (Gould 2000).

Often, as in the case of those opposing GMOs, true believers will ignore the mass of cumulated evidence in the form of field research, experiments, and numerous peer-reviewed articles and scholarly books that can find no evidence of harm in that which they oppose, but expect a single report of any kind or quality that purports to show harm, to close out all further inquiry. Peer review is an imperfect process and, like science and the scientific method, it is a work in progress. Recently, the peer-review process in medical research was itself peer-reviewed and found wanting (JAMA 2002). Before that, some of the leading journals in science had come together to set stricter guidelines for accepting papers and requiring authors to state any possible conflict of interest and not to publish any funded research in which those engaged in it were in any way restricted or prohibited from publishing negative results. However the rules may have been abused in the past and however the reforms may be imperfect, open inquiry, the scientific method, and peer review are self-correcting processes. Those who wish to criticize them, not to reform them but to overturn them, have yet to offer anything even remotely as effective in expanding human knowledge and human capability to expand the potential of the human life process. Nor are they willing to disclose fully their ties to funding from interests, including those selling organic products, that benefit from their advocacy.

NGOs and Protest

Some of the NGOs leading the protest and claiming the moral high ground have annual budgets in excess of $100 million and have yet to do anything to help the poor or anyone else feed themselves. Other NGOs carry names with food or rural in them but their accomplishments in agriculture or rural development, if any, pale in comparison with their rhetoric. To the extent that they are able to obtain aid agency funding (the Scandinavian aid agencies have been particularly generous to them) that might have otherwise used them for development, one might argue that their net impact has been negative even apart from their efforts in opposing the new technologies that offer the best hope that we have of feeding the expected nine billion people. The best estimates are that the percentage of "total income for development NGOs derived from donors" had risen by the mid-1990s "to around 30 per cent" (Hulme and Edwards 1997, 6–7). For some countries, the percentage terms are far greater—United States: 66 percent, Canada: 77 percent, and Sweden: 85 percent as "dependency ratios of between 50 and 90 per cent are common" (Hulme and Edwards 1997, 21). For 1993–94 government funding for NGOs, "the OECD estimate of US$5.7 billion is certainly an underestimate by as much as US$3 billion according to one World Bank estimate" (Hulme and Edwards

1997, 6). In a considerable understatement, the authors conclude that "channeling funds to NGOs is big business." This is ironic for groups that refer to themselves as "civil society."

Jose Bove has been a central figure in the annual NGO social forums in Porto Alegre, Brazil, which claim to represent those in need such as poor peasant farmers. The governor of the state who allowed himself to be identified with Bove, not only failed to win reelection but also failed to win his party's primary in Rio Grande Do Sul owing to peasant farmers who saw Bove's agricultural protectionism as a threat to their livelihood (Rohter 2003). It is noteworthy that when the poor can speak for themselves, their message is at variance, often exactly opposite, to that of those who presume to speak for them as Bove does.

Never have I seen an issue where such a preponderance of scientists with knowledge and experience was so solidly on one side of an issue, as is the case for the intelligent use of biotechnology in agriculture and pharmaceuticals. Those opposed to biotechnology have garnered the vast majority of the publicity in needlessly frightening people about what has been deemed by the National Academy of Sciences, the Royal Society, and others as the safest, most predictable form of plant breeding yet known.

Feeding Nine Billion

In the new millennium, feeding the hungry has been added to the environmental NGO agenda, as if the antitechnology rhetoric of past Earth Days when the poor were forgotten can accommodate this newly discovered concern. Since the first Earth Day, the planet has added two billion people, bringing the population to six billion who are living longer, are better fed, and are in better health than ever before. A look at some of the changes of the past two centuries might give us guidance as to how to both further the goals of environmental action and feed everyone in the next half century when the world is expected to add another three billion people before leveling off or even declining.

In 1800, the world's population had not yet reached one billion; by 1900, the population was about 1.7 billion. Since 1900, world population has increased better than three and a half times. The population of the United States has quadrupled, yet we produce vastly greater quantities of food on less land than was cultivated in 1930. World population has more than doubled since 1960, yet per capita caloric consumption has gone up even faster. Per capita caloric intake in developing countries has increased from about 1900 calories per day in 1960 to about 2700 today. All this has been achieved even though the land under cultivation in the world has increased from 1.4 billion hectares in 1960 to 1.5 billion hectares today. The percentage of the world population in hunger and poverty has fallen in absolute percentage terms from 50 percent in 1950 to 30 percent in 1970 to a still unacceptable 19 percent today. Though there are some uncertainties, the best estimates show a decline of two hundred million in the number of people living on less than a $1 a day over the last two decades of the twentieth century even though world population grew by 1.6 billion (Deaton 2002, 4). The belief that the world's distribution of income has been growing more unequal has been the driving force for antiglobalization demonstrations, which would lose their raison d'être without it. The evidence is overwhelming that this is not the case (Bhalla 2002). The world has gone from having a bimodal income distribution in 1950 to having a global middle-income group with the rapid rise of income in Asia (Sala-i-Martin 2002).

Stephen Budiansky argues that affluence may be the best way to preserve the environment (Budiansky 2002a,b). "Even with its vastly greater amount of meat produced and its large exports, the U.S. uses less total agricultural land (arable plus grazing) per person than does sub-Saharan Africa" (Budiansky 2002a, 33). "As countries become wealthier, yields increase substantially. . . . Use of chemical fertilizer and hybrid and other high-yielding varieties of grains could let developing countries match Western diets with little or no increase in land use" (Budiansky 2002b, 581).

Budiansky adds data showing that U.S. population more than doubled in the last half of the twentieth century while agricultural land use remained constant because of even greater increases in yields. As per capita food production, domestic consumption, and exports rose, agricultural land per capita fell. "Paul Ehrlich has argued that the environmental impact is proportional to population times affluence. But as far as land requirements are concerned, it appears to be proportional to population divided by affluence" (Budiansky 2002b, 581).

One might not call the Haber-Bosch synthesis of nitrogen fertilizer the greatest invention of the twentieth century as Vaclav Smil has done, but it would be difficult to argue against him, as we simply could not have fed even half the world's population today without it. Clearly the increases in yield from the green revolution technologies were necessary to feed earth's growing population. Can one imagine the ecological devastation that would have resulted without yield increases if the world's population had to be fed by bringing forest and wildlife preserves, scrub, mountainsides, and other marginal lands under cultivation?

The forecasts of mass famine of the 1960s and since by many who now oppose agricultural biotechnology would have come true without the very yield increases brought about by the very modern agronomy they have consistently opposed. Before the introduction of modern chemical pesticides, the origin of organic agriculture was in opposition to the use of minerals and synthetic fertilizers in crop production instead of manure. Back then, pesticides in use included various arsenic, copper, and sulfur compounds, many of which are still approved for use in organic agriculture. It is an article of faith among many of academic postmodernists, environmentalists, and ecofeminists that the green revolution was a failure though no mention is made as to how we would feed today's world population without it.

Modern agricultural technology has met the challenge of a rapidly growing world population though much remains to be done to eliminate hunger and malnutrition as well as meeting the needs of the three billion increase in population in the next half century. Most of the world's most productive lands are already under cultivation and there are increasing concerns about erosion and water contamination. Without a continuing flow of new technology, forests and wildlife preserves could be lost to agricultural expansion with the ever increasing possibility of species extinction and consequent loss of biodiversity.

No matter how they phrase their critique of the green revolution and its failure to feed people, the critics can not avoid the implication that farmers are stupid. In the high-yielding varieties of rice alone, the farmers who have adopted the green revolution package and have been planting high-yielding varieties for decades number into the hundreds of millions, producing enough additional output from increased yields alone to feed well over a billion people (DeGregori 1987). With hundreds of millions of farmers in many different crops all over the world, what forms of coercion could have forced them into this form of agricultural activity and then forced them to continue in a practice so contrary to their best interests? Are they simply stupid and need northern-white-male-dominated NGOs to show them the light and protect them from their own ignorance?

In the brief time since plant geneticists developed and won approval for the first transgenic plants, biotech crops have been adopted and cultivated by farmers in the United States and around the world. Using them, farmers are conserving agricultural resources and producing high yields with less environmental stress. In spite of the successful efforts of NGOs to frighten consumers around the world, threatening their markets, farmers are still finding it beneficial to plant transgenic crops. The hectares in transgenic crops and the new varieties are increasing every year and are likely to do so indefinitely into the future. And we are all better off because of that.

Plant biotechnology is not simply a luxury but increasingly a necessity. Though rice yields have tripled over the last thirty years, we are now "fast approaching a theoretical limit set by the crop's efficiency in harvesting sunlight and using its energy to make carbohydrates" (Surridge 2002, 576). According to John Sheehy, plant ecologist at IRRI, "the only way to increase yields and reduce the use of nitrogen fertilizers is to increase photosynthetic efficiency" (quoted in Surridge 2002, 577). Plant evolution has shown us an improved pathway for photosynthesis. "On at least 30 separate occasions, different plant lineages have evolved to use the Sun's energy more efficiently, making sugars in a two-stage process known as C4 photosynthesis" (Surridge 2002, 578). Surridge adds: "About 10 million years ago, falling concentrations of carbon dioxide in the atmosphere gave plants using C4 photosynthesis an important selective advantage. The ancestors of maize were among these plants" (Surridge 2002, 578).

Rice, wheat, and most other cereals use conventional C3 photosynthesis. The need in agricultural plant breeding is for a variety of different types of research technologies including biotechnology as well as the technologies of longer standing, which brought us to where we are today (Powell 2002; Terada et al. 2002). The sequencing of the genome of two varieties of rice will be an important new tool in creating rice varieties with genes that express the C4 enzyme (Ronald and Leung 2002; Goff et al. 2002; Yu et al. 2002). It is also likely to provide valuable insight for work on wheat, maize, and other grains which, along with rice, provide two-thirds of the world's calories (Cantrel and Reeves 2002; Serageldin 2002). Biotechnology engineering in iron-rich rice is likely to be an important factor in "fighting iron deficiency anemia," which affects about "30% of the world's population," mostly women, and is the important nutritional deficiency

(Lucca, Hurrell, and Potrykus 2002). The American College of Nutrition had a special supplement in its *Journal of the American College of Nutrition* titled *The Future of Food and Nutrition With Biotechnology* with an excellent series of scholarly articles on the many potential nutritional and other health benefits to GM food (Grusak 2002; Harlander 2002; Liu, Singh, and Green 2002; Lonnerdal 2002; Korban et al. 2002; Rocheford et al. 2002).

Improving the photosynthetic efficiency of rice has the potential of increasing nutritional value and enhancing its ability to withstand environmental stress. Harnessing of solar energy

> by photosynthesis depends on a safety valve that effectively eliminates hazardous excess energy and prevents oxidative damage to the plant cells. Many of the compounds that protect plant cells also protect human cells. Improving plant resistance to stress may thus have the beneficial side effect of also improving the nutritional quality of plants in the human diet. The pathways that synthesize these compounds are becoming amenable to genetic manipulation, which may yield benefits as widespread as improved plant stress tolerance and improved human physical and mental health (Demmig-Adams and Adams 2002).

Demmig-Adams and Adams add that terms like vitamins, "antioxidants, and phytochemicals are not mutually exclusive. Major groups of phytochemicals (produced by photosynthetic organisms) include isoprenoids, phenolic compounds, sulfur compounds, and essential fatty acids. . . . Enhancing the photosynthesizers' own protective systems may also improve the nutritional quality of foods, because fundamental cellular signaling processes and protective mechanisms are highly conserved" (Demmig-Adams and Adams 2002).

Photosynthesis involves "collection of solar energy and its efficient conversion into chemical energy," a process susceptible "to damage by any excess solar energy." Because of the "parallel functions of antioxidants in plants and humans, new mechanistic hypotheses should incorporate information from both plant physiology and human physiology" (Demmig-Adams and Adams 2002). "Protecting photosynthesis in the face of environmental stress as well as protecting human health against environmental or pathological stress requires improved understanding of molecular functions and the inter-section between stress, disease, and physiology for both plants and humans" (Demmig-Adams and Adams 2002).

Informed, intelligent criticism is essential to keep agricultural research operating to the benefit of all humankind. Opposition based on clever slogans and misinformation can drown out the voices of those with legitimate concerns who might be hesitant to speak out to avoid being identified with those whose knowledge and agenda are suspect. Critics of biotechnology would gain greater credibility if they were better informed and could demonstrate any substantial experience in helping to feed people.

CHAPTER 10

Romantics and Reactionaries

Romanticizing the Past

Too often, those who romanticize other cultures or earlier peoples are extremely selective in the data they use or in what they borrow from an intellectual discipline other than their own. A common recent practice is to take the highest estimates of pre-Columbian Indian population in order to demonstrate a larger diminution of the population following contact and conquest by Europeans. This is clearly not objectionable as long as the reader knows that there are other equally reputable population estimates that give a different picture of subsequent events.

Even more egregious than the above misunderstanding of some of the basic elements of historical demography and scholarly inquiry is the lack of understanding of the laws of conservation of matter and energy by a Third World physicist who is displayed around the world by developed countries' environmental groups to simulate support for their causes that is otherwise lacking in developing countries. Dr. Vandana Shiva, in a book-length diatribe against the green revolution, frequently refers to its voracious demand for chemical fertilizers and indicates that there were alternative ways, more benign, of achieving these outputs (Shiva 1991). Plants need ingredients (nutrients) in order to grow. If a molecule is in the plant, it or its constituent elements must come from somewhere. Except for carbon dioxide from the atmosphere, plants derive their nutrients from the soil or, in the case of nitrogen, from atmospheric nitrogen mediated by cyanobacteria. More plant output means more nutrient input. If sufficient nutrient is not in the soil, it must be added. Shiva's claim that one can grow plants without nutrients or that one can achieve the same output as green revolution seeds yield without providing nutrient input is patently nonsensical and violates our fundamental knowledge of physics.

Even before the green revolution dramatically increased the demand for and use of synthetic fertilizer, there was a large difference between the nutrients extracted from the soil in India and the so-called organic nutrients available to be returned to it. In the 1960s, each year cultivated crops in India were removing "3 million tonnes of nitrogen, 1.5 million tonnes of phosphorus oxide and 3.5 million tonnes of potash . . . 8 million tonnes of plant food. The organic sources of the plant food returned to the soil is hardly 1.8 million tonnes of nitrogen, 0.60 tonnes of phosphorus oxide and 1.8 million tonnes of potash . . . 4.2 million tonnes of plant food" (Randhawa 1983, vol. 3, 314–17, using data from Agarwal 1965, 7, 12, 13, 14, 214).

Randhawa adds: "Even allowing for the biological and other natural processes for recuperation of fertility, the balance is tremendous" (Randhawa 1983, 317). Nearly twice as much nutrient was being withdrawn from the soil than was being returned. Contrary to Shiva's assertions, this process was not sustainable. Given the dramatic increases in Indian agricultural output over the last four decades (which more than accommodated a doubling of population), the deficit in organic nutrient must be vastly greater today. Shiva cites Sir Albert Howard (whom Shiva calls Alfred) whose vitalist ideas on organic agriculture were developed in colonial India (Howard 1940). Even though he was a strong advocate of composting and had developed improved methods for creating compost (Indore method), Howard recognized the need for additional synthetic fertilizer and improved seeds. The Indian agricultural historian M. S. Randhawa, after praising Howard, refers to him as a "crusader" and warns of the danger of entrusting "policy matters to people with single track minds" (1983, 317).

Shiva has an unverified belief that "food crops for local needs" are "water prudent" (Shiva 2000). For the green revolution grains, the primary output is a larger percentage of the plant (harvest index) and therefore requires *less* nutrient input per unit of output. Biotechnologists are working to create even more efficient plants, including in the use of water, all of which is opposed by Shiva and her followers. In her paeans in praise of cowdung, Shiva's pre-green revolution Indian agriculture is one of a healthy, self-sufficient, calorically adequate, nutritious food supply produced in an ecologically sustainable manner (Avery 2000). Why hundreds of millions of peasant agriculturalists would voluntarily abandon this utopian existence for modern agricultural technologies is never explained. Equally unexplained is why life expectancies have risen so dramatically throughout Asia for both rural and urban populations, including India, if as Shiva argues, modern technology is pauperizing them and in many cases driving them to suicide. Even more difficult to explain is why those in developed countries, who are presumed to be educated and informed, uncritically accept her musings and pay her homage, including selecting her to give prestigious presentations such as the Reith Lecture (Shiva 2000; Scruton 2000).

Contrary to the claims of Shiva and others about the green revolution's voracious water use, in agriculture "water productivity increased by at least 100 percent between 1961 and 2001" (FAO 2003, 25). The major factor behind "this growth has been yield increase. For many crops, the yield increase has occurred without increased water consumption, and sometimes with even less water given the increase in the harvesting index" (FAO 2003, 25).

For wheat and rice, two major crops of the green revolution, "water consumption experienced little if any variation during these years" as per capita water use in food production fell by half (FAO 2003, 25). FAO argues that genetically engineered crops can contribute to increased "water use efficiency" (2003, 28).

Modern conservation tillage (or reduced, minimum, or no-tillage) agriculture using pesticides for weed and pest control conserves water, soil, and biodiversity better than its organic competitors and better than any previous forms of tillage (DeGregori 1985, 111–12). Conservation tillage is building up soil and soil quality. Planting with a drill, possibly disking the field, preserves soil structure and vegetative cover (and the diversity of life therein) and preserves the earthworms and other life-forms that are often destroyed by deep plowing as used in organic and older forms of conventional agriculture. These practices have been expanded in recent years with crops genetically engineered for pest resistance or for herbicide tolerance, which allows forms of conservation tillage in which a less toxic broad-spectrum pesticide is substituted for multiple sprayings of an array of targeted pesticides and herbicides thereby reducing overall pesticide use.

Many like Shiva who are promoted in the West by the greens and win uncritical acclaim are often the object of very severe criticism in their own countries, a fact that is largely unreported. After an article in a Malaysian newspaper on Shiva replete with fulsome praise and claims that she was a leader of the famed Chipko (tree huggers) movement in India, the activists in the movement sent a letter of protest to the editor. "The interview is based on false claims of Vandana Shiva and has angered many. . . . The real activists are so simple that they do not know why Vandana Shiva is reportedly publishing wrong claims about Chipko in the foreign press. We should all stand up against this new green exploitation of the people's simplicity and courage by clever, greedy and selfish persons like Vandana Shiva" (Jardhari 1996, quoted in Bandyopadhyay 1999).

Shiva uses Chipko as a model for green ideologies from deep ecology to ecofeminism. Jayanta Bandyopadhyay, a distinguished scientist, forester, and environmentalist, examines each of these in turn and deems them myths without any basis in fact (1999). He is an active supporter of the Chipko villages, which he finds "a movement rooted in economic conflicts over mountain forests" and a "social movement based on gender collaboration" and not a "feminist movement based on gender conflicts" (Bandyopadhyay 1999). His purpose is to understand and support the movement and not exploit it to promote an ideological agenda as Shiva does.

The original motivation for "participating in Chipko protests" was for local control of forest resources in order to create a "forest-based industry," which "offered the possibility that their kinsmen" who had to migrate to find work "might be employed closer to home." Further, increased local access to forest resources "may have offered women the possibility of adding to their meagre incomes and insuring themselves from potential crisis if remittances ceased or became intermittent" (Rangan 2000, 199–200).

Chipko is yet another example of developed country environmental groups coopting a cause like wildlife or habitat conservation or a local movement with legitimate grievances and subverting them. In the case of Chipko, the cooption was initially by urban elites in India like Vandana Shiva who used it to gain international acclaim. As with other cases that we examine in places like Africa and the Americas, local concerns that may have initiated the movement get brushed aside and the locals often are worse off because of the external "support" (see chapter 11 of this volume and DeGregori 2002a, chapter 2). "One of Shiva's 'Chipko women' from the Pindar Valley in Chamoli District, Gayatri Devi, bitterly states that the movement has made life worse in the valley: 'Now they tell me that because of Chipko the road cannot be built [to her village], because everything has become *parovarian* [environment]. . . . We cannot get even wood to build a house . . . *our ha-haycock* [rights and concessions] *have been snatched away'"* (Rangan 2000, 42).

This helps to answer the questions that Rangan raises: "Why do words like *environment* and *ecology* make so many people living in the Garhwal Himalayas see red? Why do so many of them make derisive comments when the Chipko movement figures in any discussion? Why is it that in most parts of Garhwal today, local populations are angry and resentful of being held hostage by Chipko, an environmental movement of their own making?" (Rangan 1993, 155). When the world community was ready to hear the claims of the Garhwal Himalayan villages, "their voice in the Chipko movement had all but ceased to exist. The brief love affair between Chipko's activists and the state had resulted in the romantic ideal that the Himalayan environment by itself mattered more than the people who eked out their existence within it." Rangan adds that "if some of the communities are ready to banish their axes today, it must be seen as yet another attempt to affirm themselves and give voice to the difficulties of sustaining livelihoods within their localities" (174–75). From Agarwal and Narain, we learn that the situation has driven some to advocate practices that violate laws that the urban conservationists have imposed. "Uttarkhand, the land which gave birth to the Chipko movement, now even has a *Jungle Kato Andolan* (cut the forest movement). Thanks to the ministry of environment, 'environment' is no longer a nice word in Uttarkhand" (1991).

Rangan argues that the Chipko today is a "fairy tale": "a myth sustained and propagated by a few self-appointed spokespeople through conferences, books, and journal articles that eulogize it as a social movement, peasant movement, environmental movement, women's movement, Ghandian movement—in short, an all-encompassing movement, beyond compare" (Rangan 1993, 158).

Chipko is a recognized "transregional environmental movement within India." It has "overwhelmed the local struggles over access to forest resources. The environmental rhetoric of saving 'nature' in the Himalayas took over, and Chipko today has been reduced to a 'green' symbol that conjures up images of bucolic mountain folk hugging trees" (Friedmann and Rangan 1993, 9).

Contradictions do not seem to bother Shiva and those who revere her. In the same public lecture in Toronto, Canada, before an adoring audience, Shiva has the price level of food in India doubling and falling

at the same time. Arguing the "failure" of the green revolution technologies, she has the price of food in India doubling so that consumers
can no longer afford it. When she wishes to criticize the United States
for "dumping" food on the Indian market, pushing Indian farmers to
commit suicide, she has subsidized foreign food "driving down prices"
(O'Hara 2000; Oakley 2000). The following excerpt from a news item
on Shiva's visit to Houston in October of 2000 is indicative. Shiva appears not to know the difference between a field of rice and one of
weeds. Such is her expertise. Shiva is touted as a "visionary, author, activist and eco-feminist." "Shiva walked across the road and looked out
into a shaggy field. 'They look unhappy,' she said. 'The rice plants.
Ours at home look very happy.' 'That,' RiceTec reports, 'is because it's
not rice. That's our test field, it was harvested in August. That's
weeds'" (Tyer 2000).

Cooking and Eating Closer to Nature

Not only is natural not necessarily better, food processing is an essential component of food safety. Many of the important transitions
that humans made in food production, particularly for grains but also
for tubers, depended on food processing for their full realization. Food
processing has been historically a vital part of human development.
The use of fire for cooking allowed our progenitors to make use of a
vastly greater array of nutrient sources by cooking vegetables (Bittman
1994). "Some cultivars are quite toxic, unless properly prepared"
(Garn 1994, 90). We often speak of the various nutrients in the foods
we eat but any nutritionist will verify that there is a difference between
the various nutrients in food and the ability to utilize them. Nutritionists will speak of "bioavailability" to indicate what nutrients in a foodstuff are available for human nutrition. Some of the most important
plant research being carried out in food crops is simply to suppress the
mechanisms in a plant which prevent utilization of an important nutrient.

Food processing has been a necessary component for the use of
plants and animals that became the basis for the population density
necessary for establishing civilization. "Man's vegetable diet, without
fire for cooking, is pretty much limited to special plant products like
fruits and nuts" (Bates 1967, 39). "Cooked foods require less extensive
digestion than raw plant foods, the adoption of cooking can influence

the morphology of dentition and the intestine, reducing tooth size and gut size" (Wrangham et al. 1999, 568; see also Milton 1999, 583; Susman 1987; Aiello and Wheeler 1995; Brace 1996). There was, then, "a reduction in the muscles of mastication and in tooth size and tooth complexity, resulting in changes in the entire craniofacial complex" (Milton 1999, 583; see also Armelagos et al. 1984; Carlson and Van Gerven 1977). Our "digestive tract is simply not equipped to deal with cellulose and raw starch, which make up the bulk of vegetable material." The "cellulose walls of plant cells are broken down by heat, and the starch is chemically changed into more digestible forms" (Bates 1967, 39). Because of cooking, "we don't need large fermentation chambers to break down long-chain carbohydrates" (Angier 2002). Cooking has been described as a type of "external, partial predigestion" (Bates 1967, 39).

Eating closer to nature has become the latest imperative of the food faddist. This means eating food raw (not irradiated) whenever possible, which carries considerable risks. One pundit maintains that "if you want to be healthy, you have to stay as close to nature as possible. When an animal or plant is raised in an organic environment, you are giving your body what Mother Nature intended" (quoted in Pence 2002, 6–7). Whatever nature intended, our biological endowment tells us otherwise as we have seen concerning our dentition and digestive tract. These changes in physiology were part of a process that produced larger brains and erect posture, freeing the hands for tool making.

An interesting and controversial theory has the use of fire and cooking playing a central role in the in the biological and social evolution of our hominid ancestors (Wrangham et al. 1999; Wrangham 2001). The details of the controversy, which largely involves the dating of the first controlled use of fire, need not concern us as both the proponents and critics of the theory largely agree on the important role cooking and other forms of food preparation have played in regularizing food supply, making more foods accessible (making more of the living matter of the environment food for humans), unlocking more of the nutrients in the food that was eaten, and making it safer. "The ability to cook food opened a very large niche for people, allowing them to eat foods they couldn't eat before, and to be in places they couldn't exploit before" (James O'Connell, quoted in Angier 2002). "Cooking makes food more available and digestible by (1) cracking open or otherwise destroying physical barriers such as thick skins or husks, (2) bursting cells, thereby making cell contents more easily available for digestion or absorption, (3) modifying the three-dimensional structure of molecules such as proteins and starches into forms more accessible for digestion by enzymatic degradation, (4) reducing the chemical structure of indigestible molecules into smaller forms that can be fermented more rapidly and completely, and (5) denaturing toxins or digestion-reducing compounds" (Wrangham et al. 1999, 570).

To get the same energy from uncooked food, a person would have to eat twice as much on a vegetarian diet and 50 percent more on a diet of meat and plants than would be necessary if they cooked it. "Adding raw meat to the diet doesn't help much because it takes a long time to chew." In a modern study of raw food eaters in Germany, about one-third "suffered from chronic energy deficiency and half of the women did not have a regular menstrual cycle" (Randerson 2003, 37; see also Wrangham and Conklin-Britain 2003). Angier adds that "not only does cooking make food delicious, it also makes it safer and more digestible, the better to extract the maximum number of calories from any given meal." Angier sums up the argument as follows: "Cooking bursts open the cells of foods and releases their nutritious innards; it breaks down long, tough chains of proteins and carbohydrates into simpler and more digestible sugars and peptides; and it detoxifies many of the defensive and potentially sickening compounds in plants, as well as killing many dangerous meat-borne microbes" (Angier 2002).

Regularized food supply has always been an important factor in species survival and has been a major contributor to longer life and good health for humans in modern times. Though on a different scale than the present, the development of cooking contributed to the regularized food supply of the first humans who used it. For early humans, seasonal shortages meant that "preferred foods such as fruits and seeds would not have been consistently available and dental and ecological considerations" did not allow the use of "herbaceous leaves and piths that make up the fallback foods of modern African apes such as chimpanzees" (Wrangham et al. 1999, 570). Plant roots and tubers constitute a form of "underground storage" of nutrients to which stone tools and digging sticks allowed humans to gain ready access and cooking made them palatable. "Cooked foods available for infants would also have enabled mothers to shorten the period of weaning" (Wrangham et al. 1999, 577; see also Hawkes, O'Connell, and Blurton Jones 1991).

As we have previously noted, there is substantial literature in anthropology arguing that high-caloric-density foods such as meat were essential for the emergence of humans (DeGregori 2001, 77–81). Scavenging has been seen as the pathway that allowed our hominid ancestors to increase the meat in their diet and join "the large carnivore guild" (Walker 1984, 144; quoted in Brace 1999, 577–78). When our ancestors moved out of Africa into the colder northern Europe, cooking became "obligatory." "Control of fire" was not only "necessary as compensation for the tropical physiology bequeathed to us by our ultimate African heritage, but it was also mandatory in ensuring access to essential edibles. . . . Mammoth hunters could scarcely have eaten a whole pachyderm in a single meal" (Brace 1999, 578). "After the lapse of a Pleistocene winter weekend, the rest would then have become frozen solid before it could have served as the basis for another repast" (Brace 1999, 578). Fire and cooking greatly expanded the range of human habitation in otherwise colder climates where there may have been abundant potential food resources, provided humans could gain access to them. Brace further argues that it is difficult to "imagine" how fire could have been "maintained and perpetuated" without language and that the necessity of controlling fire for cooking and cold climate habitation may have been critical factors in the development and use of language. Control of fire is facilitated by structured hearths and ovens (Corbey and Roebroeks 1997, 919; Wrangham 2001, 137–38).

It is almost axiomatic that the predator has to have more intelligence than the potential prey, particularly if the prey has greater quickness and speed that requires the predator to use stealth and cunning. Cooking and meat eating gave our ancestors the high-caloric-density food to provide the energy for the energy-demanding human brain and facilitated the evolution of the "increased innate intellectual capacity that accompanied the conversion of that unlikely biped into a facultative carnivore," which is "almost certainly the reason for the increased brain size" (Brace 1999, 578; on the benefits of increased consumption of meat for those whose current consumption is low, see Liua, Ikedaa, and Yamoria 2002). What "is extraordinary about our large brain is how much energy it consumes—roughly 16 times as much as muscle tissue per unit weight . . . at rest brain metabolism accounts for a whopping 20 to 25 percent of an adult human's energy needs" (Leonard 2002). Increased brain size is an "investment with initial costs and later rewards, which coevolved with increased energy allocations to survival (Kaplan and Robson 2002; see also Raichle and Gusnard 2002). Humans "produce less than they consume for about 20 years" while they use their brains to acquire the skills to produce a surplus to feed the new members of the group with a pay-off coming in a longer life (Kaplan and Robson 2002).

Cooking in the larger sense of food preparation would have included the technologies of "mortars and pestles for pounding and grinding roots and seeds, which are not easily digested by the human gut unless they have been altered by both mechanical and thermal means" (Brace 1999, 578; see also Milton 1999, 584). Agriculture became a meaningful possibility when cooking made a huge variety of vegetable resources accessible to humans as a "major basis for human sustenance" (Brace 1999, 578). Another "crucial technology" was the container "so that small items can be amalgamated for transport, processing, and storage" (McGrew 1999, 583). Even after the development of agriculture, the utilization of some foodstuffs had to await the development of even more sophisticated means of preparation. For example, "only around 3000 B.C. did the Chinese discover techniques for deactivating the anti-trypsin factor (ATF) in soybeans (which cannot be accomplished by ordinary cooking), with the result that beans and then, some 2,000 years later, bean curd could become dietary staples" (Milton 1999, 584, citing Katz 1987). "Smoky wood fires may have been used for the preservation of meat and for the oil-tanning of hides. Fire may also have been used to control animals when hunting; to burn land deliberately to increase the productivity of plant resources (and thus to attract animals); and to fell trees or hollow-out tree trunks" (Janick 2002).

The skills involved in controlling fire and cooking added other dimensions to the quality of early human life. "Charcoal was used for pigments, as has been shown by recent analysis of black paintings in the Magdalenian Salon Noir of the cave of Niaux. The frequency of remains recovered from deep caves without signs of fire suggests the use of some form of portable light source, and this is confirmed by finds of stone lamps, possibly used to burn vegetable wicks in animal fat and by the fragments of wood, probably torches, found in caves such as Basua" (Janick 2002).

The current trendy passion for uncooked food is solidly in the vitalist tradition. The raw and uncooked are so-called living foods. Eating closer to nature is not natural, whatever that means. It still deprives us of vital nutrients as it did our hominid ancestors. Even with our greater accessibility to an incredible array of foods, heat, germination, and fermentation are needed for some nutrients to be accessible. Fish is a source for the B complex vitamin and thiamine essential for metabolizing carbohydrates and for the maintenance of neural activity. Its deficiency in adults results in beriberi. Fish has the enzyme thiaminase, which inhibits utilization of the thiamine. Thiaminase is destroyed in cooking increasing the availability of the thiamine without in any way diminishing "the beneficial oils found in seafood" (CSIRI 2002). Similarly, while cooking decreases the vitamin C in tomatoes—vitamin C is generally reduced by cooking—it more than doubles the level of the antioxidant (therefore, anticarcinogen) lycopene (NNF 2000). The title of a research article sums up these findings: "Thermal processing enhances the nutritional value of tomatoes by increasing total antioxidant activity" (Dewanto, Wu, and Liu 2002; Dewanto et al. 2002). For the consumption of uncooked tomatoes, biotechnology has produced anticarcinogenic tomatoes with enhanced "phytonutrient content, juice quality, and vine life" (Mehta et al. 2002; Giovannucci et al. 2002; Mackey 2002; Sevenier et al. 2002).

Some nutritionists conclude that "some loss is inevitable, but for most, nutrient losses are small" (Davidson et al. 1979, 213). As with most human endeavor, some sense of balance is useful. Most everyone I know sometimes eats vegetables raw and sometimes eats the same vegetables cooked. In the light of what we now know and are learning, this seems to be the most sensible strategy. Eating closer to nature may be trendy and give one a sense of superiority, but nutritionally, it doesn't make sense except for the fact that modern science and technology has made the availability of nutritious food so abundant that it takes a truly extreme food fetish to cause substantive harm.

In California, people make restaurant reservations a month or more in advance to pay $69 a person to eat uncooked food while there are uncooked "potlucks in Little Rock, festivals in Portland, conferences in Boston, tropical retreats in Bali. A small library's worth of 'uncookbooks' have been published." Orenstein notes "a growing number of people who believe that eating uncooked 'living foods' extends youth and staves off disease—who, in some cases, consider cooked food tantamount to poison. Heat, they maintain, depletes food's protein and vitamin content" (Orenstein 2002).

Some believe a living foods diet "delays aging, boosts energy and prevents or cures virtually all life-threatening diseases" (Orenstein 2002). To one author, "we can't control terrorism, but we can make sure we don't eat anything cooked" (quoted in Orenstein 2002). "Raw-food rules allow heating to 118 degrees," which is a marvelous temperature to multiply most microorganisms. The "cuisine may be raw, but it is not unprocessed." Orenstein says, "one paradox of uncooking is that it is more labor intensive than cooking. And it is heavy on the gear: the must-haves include a hydraulic juicer and colossus-size food processors" (Orenstein 2002).

Modern Food Supply and Safety

Food contamination has been a fact of human existence. As carriers of botulism or ergot and aflatoxin from the fungus *Aspergillus flavus*, the food necessary to sustain life has caused mass illness, blindness, and large-scale death (Matossian 1989). The aflatoxins are still scourge in poor areas such as West Africa where there is evidence that fungal infested food is stunting the growth of children (Gong et al. 2002; IITA 2002). The aflatoxins have long been known to be carcinogens and to suppress the immune system. Even food uncontaminated by microorganisms contains substances that would be considered a threat to human life if they were used as a food additive. The production of toxins by plants was an evolutionary adaptation to avoid being eaten.

The term *chemicals* has become a code word for manufactured chemicals and is used to condemn food additives. Plants are also chemical factories and generate toxins in far greater abundance than the small quantities of manufactured chemicals applied to them for pest control. Some of these chemicals are for medicinal uses and some are poisons. And some of these same natural chemicals have been used for both purposes, depending upon the mode of usage, particularly the dosage—dose makes the poison. Many of the so-called naturally produced chemicals are highly toxic and have very active properties (Dale, Clarke, and Fontes 2002). "The use of toxins by plants for defense against pests and diseases is a common phenomenon in nature. Some natural defense substances can be highly toxic, such as glycoalkaloids in deadly nightshade (*Atropa belladonna*) and ricin, found in the castor bean" (*Ricinus communis*) (Dale et al. 2002).

Yield, whether it be per land or per labor unit, has been a primary consideration for food production; nevertheless, other considerations in selectivity make domesticated plant evolution a complement to processing in making food accessible to humans. Such foods can in no meaningful way be called natural.

Food and Microbes

Irrigation water or manure can be a vehicle for infecting a plant with *E. coli* 0157:H7, which can be taken up through the roots. It is difficult to wash out even with chlorine in a plant like lettuce where the *E. coli* has penetrated the "stomata and junction zones of cut lettuce leaves becoming entrapped" below the cut surface (Solomon, Yaron, and Matthews 2002, 397; see also Solomon, Potenski, and Matthews 2002; Hilborn et al. 1999; Clarke 2002). Those who prefer produce grown organically using only manure instead of synthetic fertilizer should take extra precautions when preparing a salad, particularly since they are opposed to irradiation of produce, which is about the only truly effective way of killing entrapped harmful microorganisms.

Epidemic of Cancer?

The belief that we are experiencing an epidemic of cancer or an upsurge of cancers, has become so widespread as to become common knowledge. It can be asserted without any evidence being in support of that proposition. The statistical data by the best epidemiologists are at considerable variance with this belief. As early as 1981, Doll and Peto were finding evidence of declining age-adjusted cancer rates if lung cancer caused by smoking—a lifestyle choice—was excluded (Doll and Peto 1981). At that time there was a slight increase in cancer rates among the older population, which was offset by the declining incidence among the younger cohorts. The increased incidence in the older population was explained by the fact that Medicare had significantly increased the access to medical care—diagnosis and treatment—and not by any actual increase in the rate. Screening for a disease will almost inevitably increase the measured rate by turning up additional cases, which may have otherwise been unreported. This was almost certainly the case in the breast cancer surveys that claim to find cancer clusters, which later more rigorous research find not to be the case. By the early 1990s, the evidence was emerging very strongly that there was no evidence of any "cancer epidemics" as age-adjusted data were finding declines for cancers not related to smoking (Doll, 1992; Coggon and Inskip 1994). There is evidence of a cancer epidemic related to smoking but even here on the "basis of current trends the epidemic can be expected to decline first in men and later in women" as the number of smokers (in developed countries) continues to decline (Coggon and Inskip 1994). "There is no evidence that toxic hazards such as pesticides, chemical waste, and other forms of industrial pollution have had a major impact on overall rates of cancer" (Coggon and Inskip 1994).

The End of Vitalism?

Some form of vitalism was a very reasonable belief to hold until nineteenth-century science found it wanting. Unfortunately, vitalism is not only alive today but in various forms of pseudoscience, its assumptions permeate popular perceptions to the degree that they simply trump all evidence to the contrary. Manure has a vital quality; synthetic fertilizer is dead. Organic vegetables are better for you because of a vital quality. Holistic medicine uses the vital forces for healing; modern "reductionist" medicine fails to treat the whole person.

The natural is endowed with vital properties that are lacking in the synthetic or artificial. Those who demonize DDT or transgenic food are dealing with a public's set of preconceptions fostering an uncritical acceptance of assertions about the dangers of modern life. The 1950s movie *The Invasion of the Body Snatchers* provides a template for vitalist beliefs. The pods who took over human bodies were absolutely identical in every respect to the original persons except for a vital ingredient, in this case, emotions. Vitalism argues there are vital qualities even if they can't be measured or otherwise observed. With this predisposition, is it any wonder that there is a large public sentiment believing that what is natural is safe until proved harmful, while what is created by people is harmful until proved safe? Since science can not provide absolute certainty of safety, campaigns of demonization will always be difficult to counter.

Life Expectancy Since 1840

In the many claims about the dangers of modern life and the threat to life and health by chemicals, concern is often expressed about women's health and cancer. If modern life is killing women, the statistics are clearly hiding that fact. "Female life expectancy in the record-holding country has risen for 160 years at a steady pace of almost 3 months per year. In 1840 the record was held by Swedish women, who lived on average a little more than 45 years. Among nations today, the longest expectation of life—almost 85 years—is enjoyed by Japanese women" (Oeppen and Vaupel 2002).

Oeppen and Vaupel continue that this "four-decade increase in life expectancy in 16 decades is so extraordinarily linear that it may be the most remarkable regularity of mass endeavor ever observed." The "life expectancy has also risen linearly for men, albeit more slowly: the gap between female and male levels has grown from 2 to 6 years" (Oeppen and Vaupel 2002). To the question of whether life expectancy is "approaching its limit," the authors answer that the "evidence suggests otherwise" (Oeppen and Vaupel 2002). As life expectancies have increased, experts have predicted a ceiling would be reached and, consistently, life expectancies have surpassed the expected limits (White 2002).

From the 1930s onward there was a rapid decrease in the deaths from infectious diseases particularly among young people. Many factors were involved: cleaner water, better nutrition, and immunizations. A variety of pharmaceuticals played a significant part in this reduction. "Infectious disease mortality in the U.S. declined . . . from 797 deaths per 100,000 in 1900 to 36 deaths per 100,000 in 1980" (Bailey 2002, 131). A mid-1930s survey of 700,000 households by the U.S. Public Health Service found that "an average of 51 percent of all deaths of children between 1 and 15 years of age was due to infectious and parasitic diseases, pneumonia, diarrhea enteritis" (PHS 1976, 14, quoted in Bailey 2002, 121).

Many of these deaths in the 1930s were not preventable because we lacked the knowledge and technology to control them. Today, what is killing far too many children is not modern technology and science but lack of access to it (Sachs 2001). Research presented at a meeting on environmental threats to the health of children, sponsored by the United Nations' World Health Organization, found "1.3 million small children in developing countries died from diarrheal diseases linked to unsafe water as well as poor sanitation and hygiene in 2000" (AP 2002).

The number one cause of death of children in developing areas is respiratory diseases. About "2.2 million child deaths a year were caused by acute respiratory infections associated with indoor air pollution—such as the burning of fuels for cooking and heating in homes" and other dwellings that could not afford modern, less-polluting technologies (AP 2002; UN Wire 2002). In India, "indoor air pollution caused by burning of traditional fuels like dung-cakes, wood and crop residues is causing considerable damage to the health of country's rural and semi-urban population with nearly half-a-million women and children dying each year . . . a pollutant released indoors is 1,000 times more likely to reach people's lungs than that released outdoors" (TTOI 2002). High death rates for poor children reflect large but incomplete improvements made over the last half century. "Between 1960 and 1995, life expectancy in low-income countries improved by 22 years as opposed to 9 in high-income countries. Mortality of children under 5 years of age in low-income countries has been halved since 1960" (Jha et al. 2002).

Romanticism in Social Policy

If romanticism were merely poetic license, it would be harmless. Unfortunately, it rarely is restricted to matters of the heart. Romanticizing the past has a habit of spilling over and imbuing the lifeways of prior times with an array of virtues that they simply did not have, which in modern times has led to advocating the practices of the past over the technologies and practices of the present. Much the same can be said about romanticizing the lifeways of poorer peoples who do not have the benefits of modern technologies that the more affluent of us have. However it is romanticized, seeking a return to a never existing past is a reactionary prescription no matter how radical the promoters of these policies may proclaim themselves to be. Romanticizing the lifeways of the poor is a prescription to do nothing to alleviate their condition, which is also reactionary.

In this chapter as in other chapters in the book, I give data on the improving conditions of life for those who were fortunate to be born in societies that were experiencing rapid technological change. On issues such as food preparation or overall food safety, in the name of improved health, our contemporary romantics are advocating practices that are less healthy. They want to eat their alfalfa sprouts and mung bean seeds but oppose food irradiation, which is the only way to rid them completely of dangerous "entrapped" microorganisms (Bari et al. 2003; Rajkowski, Boyd, and Thayer 2003). Romantics are opposed to the use of modern synthetic pesticides in agriculture, which means that weeds have to be removed by low-paid stoop labor in the hot sun (Fulmer 2003; Lee 2003). And now some want to get the most nutrition out of food by not cooking it, oblivious to the fact that cooking is a way of making more of the nutrients available to humans.

CORE (Congress of Racial Equality), a well-respected civil rights organization, has accused the "well-fed eco-fanatics" who "shriek 'Frankenfoods' and 'genetic pollution'" of being guilty of "eco-manslaughter" because of their "support of international policies limiting development and the expansion of technology to the developing world's poor" (Raven 2003). Romanticism as a personal choice is everyone's right but when there is the attempt to make it social policy, then it poses dangers for the rest of the community.

CHAPTER 11

Risk, Representation, and Change

Arguments against constructive change take many forms. I have called them myths of the "riskless alternative" (DeGregori 1974, 23). Every change has risk be it a political, scientific, or technological change but a simple assertion of risk is not in and of itself a valid argument against change. The risks of change have to be measured against the benefits of change and the risks of not changing. We are getting ever more impossible demands for a zero risk society. In public discourse, the demand will be for scientists to guarantee that an innovation, be it genetically modified food or a new pharmaceutical, have no possibility of ever causing harm. Given that no reputable scientist can ever answer such a question with absolute certainty, the interrogator has seemingly won the argument by default in presuming that there is some totally risk-free alternative in the status quo or in some presumed prior way of doing things. Opposition to change in favor of the status-quo-ante used to be considered a conservative or reactionary position; it has become the battle cry of presumptive radicals from the streets of Seattle to those of Genoa. Having won the argument by not being guaranteed safety with absolute certainty, believers feel no need to subject their alternatives to the same tests that might show them to carry far more risks.

Along with riskless change, there are now demands for "victimless" change. If there are possible risks, there are possible victims and in some cases, likely victims with most every change. If we examine the many changes in the past century that have done the following:

- reduced infant and child mortality over 90 percent
- added nearly thirty years of average life expectancy to American lives
- caused a rapid recent growth in disability-free years of life and comparable or greater amounts in other countries
- chlorinated water
- pasteurized milk
- regularized food supply because of synthetic fertilizers and chemical pesticides
- gave us modern medicine and immunization

we will find that all carry risks and all have had, and continue to have, organized opposition. Most every vaccine carries some risk but those we depend upon carry vastly fewer risks than the threat to life and good health from the disease that they protect against. None carry zero risk although some are getting very close to it. I ask, as before, if technology and science are killing us, why are we living so long?

Because infant and child mortality and morbidity have so successfully been reduced, we, individually and collectively forget the scourge of the diseases against which we are now protected. Infants and children still suffer from other maladies, of which we are uncertain as to the cause. Given that infants are given a successive regimen of eleven immunizations, it is likely that some will coincide with the onset of an unexplained malady. The antiscience and antitechnology coterie are quick to assign the blame to immunization, even without evidence, and frighten other parents into not immunizing their children. The evidence is overwhelming that a decline in immunization will eventually lead to an increase in the disease with death or permanent damage to lifelong health following in its wake. In the United Kingdom and Germany, these fears have led to declines in immunization, which have lowered immunization rates perilously close to the minimum for "herd immunity" and have led to epidemics of the diseases.

There is a role for the genuine radical to call attention to the victims of change. Identifying victims forces us to consider many factors; first and foremost, are there more beneficiaries of change than victims but also, are the losses fairly distributed or are some groups disproportionately gaining the benefit while others are paying the price? Focusing on victims and risks can force us continually to find ways to reduce the adverse outcomes—making our vaccines ever safer. In areas such as globalization, it is both those harmed and those who have benefited.

Critics speak of victims but not the hundreds of millions of people who have been able to rise out of poverty as a result of opening their economies to change. Identifying victims of change should be a call to action to find ways of making it fairer, sharing the benefits of it, and not be a simplistic argument against change.

The Northern NGOs and Southern Protest

There is a kind of Gresham's law of social protest where strident total opposition to a change such as globalization drowns out any more reasoned arguments for making it fairer. Wealthy advocacy groups largely controlled by white European and North American males with sophisticated command of public relations have created a new form of neocolonialism in taking control of the agenda in opposition to globalization and change. "In recent years, many social movements in conflict with developing states have sought support from trans-national NGOs. Because of contemporary developments in global communications and transportation, there are readily-available means for pursuing patrons across national borders" (Bob 2001).

Clifford Bob gives us an in-depth analysis of two of the groups, the Zapatistas rebellion of Chiapas, Mexico, and the Ogoni movement of the Niger River Delta of Nigeria (The Movement for the Survival of the Ogoni People or MOSOP), that have sought help from and been championed by the NGOs (Bob 2000, 2001, 2002). "While the pursuit of transnational support is common, gaining it is neither easy nor automatic. Most domestic movements begin in international isolation, and because many now seek external allies, they compete against each other for the limited resources of transnational supporters" (Bob 2001). The NGOs like to present themselves as moral agents of a greater cause, that of the poor and the oppressed. Bob maintains that the argument that "NGOs and transnational networks are primarily moral actors is inadequate to explain the development of transnational support. Principles create only permissive guidelines for selecting one or another needy group; they do not help us understand how 'moral' NGOs choose to support one movement rather than another" (Bob 2001). "A top-down perspective is one-sided: It ignores the active role of domestic social movements in the South in responding to opportunities on the international plane, promoting their own causes transnationally, and wooing potential NGO supporters" (Bob 2001).

The Ogoni movement was going on for several years prior to the NGOs taking notice and was concerned with "increased political power, economic justice, and control over natural resources" (Bob 2001). It was little different from many other minority groups seeking control over their lives and economic destiny and seeking some form of economic justice. Such groups exist in abundance throughout the world including Nigeria and the Niger Delta region (Ijaw, Ogbia, Ikwerre, Urhobos, and Nembe Creek communities), some of which are much larger groups with similar grievances who have been largely ignored. "None of these movements has succeeded in arousing the support of major international actors or in attracting substantial media reporting" (Bob 2001). Living in an oil-rich area and realizing few of the benefits, they were seeking their fair share of the oil revenue and not an end to it. Bob asks why was it that "the Ogoni became international *cause célèbres* in the mid-1990s, while the other Niger Delta movements did not" (Bob 2001). Bob's answer is instructive. The charismatic leader of the Ogoni movement, Ken Saro-Wiwa, sought the help of international NGOs. Saro-Wiwa made several trips to Europe to seek support. Except for the UNPO (Unrepresented Nations and Peoples Organization), the "Ogonis' earliest and most consistent NGO supporter," no help came forth from groups like Friends of the Earth or Greenpeace. Saro-Wiwa was an internationally known writer with many contacts in the developed countries. Wealthy with a large home in Surrey, England, and a son at Eton, he was the "epitome of the Anglicized product of the Empire" who "lost touch with reality as he was wooed by starry names in Europe and the US" (North 1996).

"The Ogoni Bill of Rights . . . focuses on political autonomy and portrays environmental issues as the Nigerian state's responsibility" (Bob 2001). The Ogoni had to reframe their demands to fit the ideological needs of the NGOs "to highlight environmental problems caused by a Shell subsidiary. Thereafter, Shell's 'ecological warfare' against the 'indigenous' Ogoni became an increasingly prominent part of MOSOP's rhetoric both at home and abroad" (Bob 2001). The Ogoni were now a symbol of the international struggle of Greenpeace International and other NGOs in their battle against globalization and greedy corporations.

As with other issues that we have discussed, the NGOs are particularly effective in disseminating their narrative on the Ogoni and Ken Saro-Wiwa to the virtual exclusion of all others. On certain issues, there is no disagreement. Sani Abacha, a vicious, ruthless dictator, came to power three years after the founding of the MOSOP. A succession of Nigerian governments had monopolized oil revenues denying the producing regions their fair share. The oppression by the Abacha regime and its military predecessors was not unique to these regions. Never explained is how it was in the interest of the oil-producing companies to connive with the Abacha to deny the Ogonis their rights as they had every incentive to encourage a flow of revenue into the producing regions to maintain stability in the area (Paine 1999).

There is an alternate narrative to that of the NGOs. Saro-Wiwa's adoption of the NGO rhetoric severely split the Ogoni movement and resulted in inter- and intra-ethnic conflict and killing and internal divisions in the protest movements with Saro-Wiwa's control of MOSOP slipping away. Reading the online postings of the NGOs and the Nobel Prize winning author Wole Soyinka carefully, one can verify that the movement was split though there is an attempt to blame Abacha for it (Soyinka 1995). Four moderate Ogoni leaders in Gokana Kingdom were killed reportedly by angry MOSOP youths. Many of those doing the killing were Saro-Wiwa's followers though those with whom I have discussed this matter do not believe Saro-Wiwa himself was involved in the killing. The daughter of a moderate Ogoni leader who was "lucky to escape with his life," accuses Saro-Wiwa of "incitement to murder" (North 1996). Abacha, who had no qualms about murdering those who opposed him, conducted a sham trial, convicted, and executed Saro-Wiwa and eight others. Not all of those found guilty were necessarily innocent as assumed in the NGO narrative.

The Ogoni did get the media attention that they needed and the NGOs got a powerful symbol for their ideological cause and for more effective fund raising, but it is questionable as to what the Ogoni got out of sacrificing their cause to the demands of the NGOs. And none of this has helped the other minorities with comparable grievances to have their concerns considered and addressed. To some, the actions of the NGOs would seem to be another form of neocolonialism and orientalism where an elite group in developed countries stands as gatekeeper, defining which causes are legitimate and presuming to be better able to speak for the people in developed countries than the people themselves. Even where the international support is helpful, its help is to achieve outcomes that are not entirely what the people were seeking. Further, many who might be supportive of the legitimate demands of a people are often discouraged from doing so when it becomes a symbol for other more ideological objectives.

The antiglobalization NGOs are highly selective in the biography of those whom they support. Most of those who have learned of Saro-Wiwa from the NGO narrative would be surprised to learn he was a millionaire businessman or that he was on the side of the federal government (he managed the oil port of Bonny) and not Biafra during the Nigerian Civil War and may have gained financially from it (North 1996). Good or bad, what we are told by the NGOs is what fits the image that they are seeking to create and not the larger reality, however relevant it may be. The antiglobalization NGOs are equally selective in whom they elevate to martyrdom. There were other important victims of the wrath of Nigerian military dictators contemporary with Saro-Wiwa but since they don't contribute to the antiglobalization cause, their plight is largely ignored by all but the human rights NGOs.

Millionaire businessman Bashorun Abiola was overwhelmingly elected president in 1993 in what was to many observers the fairest elections in Nigeria since independence. After having the audacity to declare himself president and taking the oath of office, he was jailed by Abacha. In 1996, Abiola's wife, Kudirat Abiola was assassinated. It is not unreasonable to assume that it was by the Abacha government. In July 1998, still a prisoner, Abiola died from mysterious causes while being interviewed by foreign diplomats. Since there were no multinational corporations or globalization to blame, only a vicious dictator, one searches in vain for the commemorations of the deaths of the Abiolas.

Dr. Beko Ransome-Kuti is a medical doctor and the leading human rights prodemocracy activist in Nigeria as chairman of the Campaign for Democracy. He is the brother of the late Afrobeat star Fela Kuti and uncle to pop star Femi Kuti. The persecution of Ransome-Kuti and his family is of long standing. In 1977, in a raid on the compound of Fela Kuti, Funmilayo Ransome-Kuti (mother of Beko and Fela and prominent political personage in her own right) was tossed out of a window and suffered injuries that proved fatal several months later. Everyone in the compound was injured including Fela Kuti whose hands were broken preventing him from playing the saxophone for several years. In 1995, Beko Ransome-Kuti was tried by the same tribunal that tried and convicted Saro-Wiwa. He was sentenced to fifteen years in prison for treason for allegedly being involved in a coup attempt. He was released in June 1998 after the death of Sani Abacha, who many believed was poisoned by his fellow officers who may have felt threatened by the international reaction to Abacha's excesses.

How many antiglobalization campaigners have even heard of Beko Ransome-Kuti? Being the brother and uncle of internationally renowned music stars, his continued harassment over the decades should have drawn more international attention from other than the human rights NGOs. A cynic might suggest that publicizing the deaths, if not murder of a winning presidential candidate and his wife, the imprisonment of a human rights activist, and the killing of his mother, the antiglobalization NGOs would undermine their attribution of globalization and multinational corporations as the primary source of evil and the cause of the death of Saro-Wiwa. Since the Nigerian government had a majority 55 percent interest in Shell Petroleum Development Company (SPDC) with Shell as operating manager holding a minority interest along with ELF and Agip, one could frame the narrative that a brutal dictator, Sani Abacha, defended a state socialist enterprise by murdering a successful capitalist businessman, Ken Saro-Wiwa. This would be a highly selective oversimplification of a complex situation but it is no more so than the widely repeated NGO litany. Given the consistency in what is included or omitted from the NGO litany, the question is, are they more concerned with defending and promoting an ideology than they are in defending those in need?

Probably more successful in using NGO-generated publicity were the Zapatistas of Chiapas, though even with a series of responses from the Mexican government, it is still not clear how many of their grievances have been addressed. "Shortly after midnight on January 1, 1994, 2,000–3,000 fighters from the Zapatista Army of National Liberation (Ejército Zapatista de Liberación Nacional, EZLN) seized four towns and the major city of San Cristóbal in the southern Mexican state of Chiapas. Previously unknown, the Zapatistas quickly drew sympathetic coverage from the international media. Within one week of the uprising, over 140 Mexican and international NGOs also surged into Chiapas" (Bob 2001).

"Their leader, the masked subcomandante Marcos" has become an international celebrity and is "articulate in Spanish and English, prolific in written and oral communication, and attuned to contemporary cultural currents" (Bob 2001). As was the case with Nigeria, there were others with grievances who did not get the publicity and support received by the Zapatistas. "Another Mexican guerrilla movement that erupted out of the oppression and poverty of neighboring Guerrero state in the summer of 1996, has failed to capture significant international support. Like the Zapatistas, the Popular Revolutionary Army

(Ejército Popular Revolucionario, EPR), voiced a vague leftist ideology; like the Zapatistas, the EPR attacked cities and government installations, this time across a wide swathe of Mexico; and like the Zapatistas, the EPR courted public opinion and the media" (Bob 2001).

Unlike the Zapatistas, the EPR failed to receive any international support or assistance and as a "result, the EPR has remained weak, the government indifferent to its demands" (Bob 2001). What distinguished the Zapatistas from EPR was that they picked up on the international NGO themes of opposition to NAFTA, international free trade and liberalization, and Indian and indigenous peoples' rights. It was no coincidence that the attack on San Cristóbal took place on the first day of NAFTA and could be used to legitimize the actions of those who opposed it. It is not very likely that the issues of a sophisticated leader like subcomandante Marcos gain international support and have much meaning to his followers. The Indians of the Lacandon rain forest are unlikely to have known much if anything about NAFTA or the policies of neoliberalism despite there being a 1996 meeting there entitled Encounter for Humanity and Against Neo-Liberalism.

In 1972, the Mexican government deeded a forest "as big as Connecticut to the tiny and untrammeled Lacandon tribe, a few hundred people, who farm by trimming the forest canopy, not erasing it." In 1978, the Montes Azules Biosphere Reserve was created expelling the Chol Indians. "Seen in satellite images the green land of the bioreserve shrinks every year, like a lake slowly going dry. The trees are cut, the undergrowth is burned, the thin topsoil planted with corn until the crop fails, the land then grazed by cattle until the rains wash the earth away. Hundreds of settlements struggle in isolation, sharing little sense of community, rarely seeing eye to eye, often lacking a common language" (Weiner 2002).

It was almost inevitable that the interests of peasants in Chiapas and many of their NGO defenders would eventually conflict as the "struggle for land has started to pit the Zapatista rebel movement against ecologists who want to save the remains of the forest" as the "movement criticizes efforts to conserve the bioreserve as a 'war of extermination against our indigenous communities.'" The conservation NGOs with funding from biotech companies speak of the "ecological dangers which the indigenous people inside the bioreserve represent" (Weiner 2002). The rebel peasant groups counter that the ecologist NGOs serve "large multinational companies dedicated to exploiting biogenetic resources" (Weiner 2002). It seems the NGOs have been hoisted on their own rhetorical petards. "'Outsiders still make the rules here,' said Ron Nigh, an anthropologist who has worked in the region for 17 years. 'Some conservationists think there shouldn't be anybody living in the jungle,' he said. 'The local people basically have no say'" (Weiner 2002).

With six billion people in close to two hundred sovereign political entities, the world is replete with legitimate grievances and groups seeking a just remedy. Tragically, they are not able to get a hearing in the media without the aid of the developed country advocacy groups who demand that they make their claims using slogans that conform to the ideology and fund-raising capability of northern groups like Greenpeace. Thus their real grievances are diluted or lost as are those of groups who have failed to use the required litany or have the integrity to refuse to do so. Justice requires that the poor and most needy have the opportunity to experience technological change without having to reframe their grievances to fit an ideological agenda of others.

We can create a world free of hunger and preventable disease in which most all its human inhabitants can live relatively long (compared to any previous era or place), productive lives except for an unfortunate few who are born with a congenital and incurable condition in terms of the medical technology of their time or whose lives are cut short by accident. We not only can do it but we have been doing it for a growing but still minority of the world's population. Even for the majority of the rest of the world's population, they are more likely to be immunized against the major childhood diseases, better fed and living longer than ever before. To achieve this future, we have to understand forces that have benefited so many as well as the forces that have worked to prevent beneficial change and are actively opposing it today. A lifelong task of mine has been to learn why we are better off than ever before and how we can work together to give others the same opportunity that those of us have who are more privileged.

Risk and Change

Reasonable people will seek to avoid unnecessary risk, which is a frame of mind with which few can argue. In avoiding risk, there is always the question as to how much anyone is willing to pay for further risk reduction. Most of us are familiar with this argument, but not as well recognized is that all of our choices in life are of risk versus risk. This is true even for the cost issues since one has to consider whether there were alternate patterns of expenditures that could have brought greater risk reduction for the same or even lower cost. Since any change not only involves the inevitable risk of every human endeavor in addition to the risk of the unknown, then opponents of change can always raise issues of risk to oppose change. What we have argued throughout is that we rarely consider the often far greater risks of not changing. What I have been describing throughout this book and in other writings is that the real and imagined risks of change are continually being raised to oppose change despite the fact that we can observe an historical process whereby change has been reducing risk and improving human life.

Today, those at greatest risk are the very poorest, particularly the poor in developing countries. That there should be global concern for their well-being and a concerted effort to help them transform their lives has become a moral imperative for many of us. The question remains of what actions are needed, and for that we need to give primacy to the expressed needs of the people themselves. Sounds simple but unfortunately it is not, because those without means are most often those without a voice. Who speaks for them? Who among them can speak on the group's behalf? Which of the many needy groups get their voices heard? I have argued in this chapter (and elsewhere) that many who claim to speak for the needy, in fact have their own agenda about which they are not always forthcoming. We have further argued that those who claim to speak for the poor are often in command of the means of obtaining international publicity for the cause of a disadvantaged group. They become gatekeepers who decide who gets heard and who doesn't and the decision is often based on the willingness of a group to refashion its grievances to fit the ideological needs and agenda of their presumed benefactors. I wonder whether some of the better known groups with very legitimate grievances have paid too high a price to have their voices heard and whether the antitechnology agenda of many of the NGOs is working against the very changes that would benefit those most in need.

Epilogue: Science, Technology, and Humanity

An item from my previous books warrants being briefly updated without more detail in the main body of the book. In *Agriculture and Modern Technology* (DeGregori 2001), I used an article that cited eight successive studies that failed to find any evidence of DDT (or other chlorinated hydrocarbons) causing breast cancer. I added a ninth that came out after the article was published and then added a tenth in *The Environment, Our Natural Resources and Modern Technology* (DeGregori 2002b). I now have an eleventh that finds no link to breast cancer from organochlorines and other environmental toxins (Gammon, Santella et al. 2002; Gammon, Wolff et al. 2002; NCI 2002). Needless to say, the advocates of the study were "disappointed" with its findings (Toy 2002). Geri Barish, the president of 1 in 9: The Long Island Breast Cancer Action Coalition, the group that success-fully sought the funding for the series of studies on breast cancer and environmental pollutants, was still certain that the pollutants caused breast cancer. Barish "knows that the pollutants studied are dangerous" and asks "How could they absolutely say that a known carcinogen is not absolutely involved in the cause of cancer?" (Kolata 2002b). Since no study will "absolutely" determine that there is "absolutely" no causal relation, then other similar studies will continue, and research resources will be consumed that could be more productively used to advance the cause of women's health.

Other studies have found that women who had fewer children had a higher incidence of breast cancer; the longer a woman breast-fed a child, the lower her risk of breast cancer (CGHFBC 2002). However politically incorrect this may sound—blaming the victim—the results should be considered as opening up a line of research that could develop pharmaceuticals or prescribe strategies for prevention. Bernardino Ramazzini (1633–1714) in 1713 observed that nuns had a higher incidence of breast cancer (they still do), which may have been the result of being childless. Women in early hunting and gathering societies or even some contemporary tribal groups may ovulate only 60 to 150 times during the course of their lifetime. Women in affluent societies may ovulate over 400 times during their lifetime. They have fewer pregnancies and more years during which ovulation and increased production of estrogen occurs because of better nutrition and longer lives. The "estrogen exposure hypothesis" argues that a woman's lifetime exposure to her own estrogen may be an important risk factor. Evolution has "designed the female body" to have more children as a survival strategy; modern life has given women longer lives, the option to have fewer children with more surviving and possibly an increased breast cancer risk (Goldstein and Goldstein 2002, 139–49; see also Byrne 2000).

After the last Long Island breast cancer study, it was revealed that the breast cancer rates there "are not much different from those of the rest of the country and a number of areas in the Northeast and elsewhere have higher rates" though figures of 30 percent higher have been the staple of activists and the media (Kolata 2002a; see also NYT 2002). The alleged source for higher rates was the New York State Department of Health and the National Cancer Institute; both deny ever making such claims. Even the lead investigator for the latest study "never alluded to a Long Island breast cancer epidemic in applying for the grant." She said that "the Long Island rates from 1987 until 1989 were 'generally comparable to the SEER rates,' referring to national rates reported by the National Cancer Institute in its Surveillance, Epidemiology, and End Results program" (Kolata 2002a).

A medical journal article on the Long Island Breast Cancer studies titled "The Epidemic That Never Was" asks the question as to whether the "activists and the media" created a "suburban legend." Quoting a journalist, the article states that "Long Island is not the breast cancer capital of the United States" but it is the "capital of breast cancer activists" (Tanne 2002).

Science, Truth, and Beauty

To Luc Ferry, the concern about romanticism and "rootedness" whether by the Nazis or romantics could be construed as the "love of nature (poorly) concealed the hatred of man" (Ferry 1995, 21–22, 28). "The hatred of the *artifice* connected with our civilization of rootlessness is also a *hatred of humans as such*. For man is the anti-natural being par excellence. This is even what distinguishes him from other beings, including those who seem the closest to him, animals" (Ferry 1999, xxvii).

Romantics do not seem to realize that all technological and scientific inquiries are simply different ways of accessing and understanding the world—nature—around us. I have often quoted the physicist Richard Feynman who complains that poets wrongly claim that science somehow "takes away from the beauty of the stars." He counters that he too sees the stars and feels them, allowing the "vastness of the heavens" to stretch his imagination (Feynman 1964, quoted in Baeyer 2000). As astronomers are probing deeper and deeper into space, this beauty and vastness is now being made accessible to more and more people.

In 1800, William Herschel observed that when light was passed through a prism, as Newton had done earlier, the spectrum was broken into different visible colors and there was an area where there was heat but no light (Rabkin 1987, 34). This was the first discovery of light waves outside the range of what is visible to humans in what we now call the electromagnetic spectrum. "Visible" light (meaning visible to the unaided human eye) constitutes only a tiny fraction of this spectrum. We have gained access to the rest of the spectrum with technology. We use it for a multitude of human purposes, in applied technologies such as X rays or radio, or aesthetic purposes in the creation of the arts, or for intellectual pursuits such as astronomy.

Since radio waves are part of the electromagnetic spectrum, we also hear parts of the spectrum and use it to transmit sounds, such as voice or music, over greater distance than we can normally hear by transforming them into electronic impulses and then transforming them back again into sound waves. For forty years, scientists have been talking about "listening" to the background noise of the big bang that took place billions of years ago. Much of our knowledge of the "universe's first 300,000 years" comes from tuning into the "cosmic microwave background radiation, the detectable afterglow of the Big Bang" (Wilford 1998). Using special instrumentation on a satellite, scientists in 1965 measured the background radiation of the universe creating "a portrait of the big bang" (Levenson 1994, 316).

More professional astronomers are studying the stars with an incredible array of instruments or visual observation, than ever before. Astronomers, amateur and professional, using the internet are conducting "extended research projects using telescopes that they have never actually visited" by connecting "underutilized telescopes to remote observations" (Ferris 1998, 58).

"Amateur astronomy—astronomy for the love of it—is enjoying a renaissance these days, in the course of which it has happily begun to unite itself with professional astronomy" (Ferris 1998, 57; see also Ferris 2002). Our affluence, high levels of education, and leisure also mean that there are many amateur astronomers (including some with PhDs in astronomy working at other professions) observing the stars each night. It has been helped by a very cheap but useful type of telescope called "Dobsonians," named for its inventor, John Dobson. "Dobson is the Martin Luther of astronomical reformation." His credo is: "To me it's not so much how big your telescope is, or how accurate your optics are, or how beautiful the pictures you take with it—it's how many people in this vast world less privileged than you have a chance through your telescope to see and understand this universe" (quoted in Ferris 1998, 57).

This is a truly democratic vision that illustrates the participatory powers of new technologies. It is essential to have both the creative vision of technological achievement and the vision of its democratic egalitarian potential. There is now a Society for Amateur Scientists that recognizes "the tremendous contributions that amateurs have made over the decades" to scientific inquiry and discovery and whose aim is to "help encourage the scientific creativity in the everyday citizen" in order to "democratize science" (Dreifus 2001).

New technologies can often be centralizing, but they don't always have to be so. Amateur astronomers with "new technological power" are doing things that could not be done by anyone just a few decades ago. Timothy Ferris has an extended list of such doings by amateurs who "record the screams of colliding galaxies, chronicle the ionized trail of meteors falling in the daytime, or listen for the signals from alien civilizations . . . observe jets of plasma arcing off the surface of the sun," while "others take photometric measurements of the cycling magnitudes of pulsating stars and eclipsing binary stars" (Ferris 1998, 59–60).

There is even an organization of amateur astronomers, TAASS or The Amateur All-Sky Survey, that has a website and an organized program of searching the sky and seeking out phenomena that are not of prime interest to professional astronomers (Chown 1998). The internet and the invention of the "light-sensing charge-coupled device, or C.C.D." have greatly facilitated the efforts of amateur astronomers. With a C.C.D. attached to a Dobsonian, an amateur has an instrument "whose light-gathering power is comparable to that of the Hale telescope at Palomar in the pre-C.C.D. era" (Ferris 1998, 58). The internet allows information and images to be shared within hours of discovery. Because these technologies allow amateurs to do so many things, Dyson finds that the "gap" separating "the amateurs and the professionals" has been "narrowed" (Dyson 1997, 73). The use of the C.C.D. in astronomy and the advance in knowledge that it has wrought is an example of what Freeman Dyson calls the "tool-driven" revolution in a science in what he calls the "digital astronomy revolution" (Dyson 1997, 66–71; Dyson 1999).

The "C.C.D. era" may well give way in a few years or sooner to CMOS chips for imaging for use in astronomy and in cameras for everyday use. CMOS is "dirt-cheap compared to CCDs" and beats them in "its low power consumption." Currently there are problems with CMOS (complementary metal-oxide semiconductor). Research is in progress to solve them so that the chips can be competitive with C.C.D. or even eventually displace them (Dvorak 1998, 89).

Possibly or even probably, there are as many people looking at the stars through telescopes today as was the case before the other instruments were invented. Thus, we have the first sighting of the supernova and the more recent sighting and naming of the Hale-Bopp comet by amateurs. Many of the discoveries of amateurs, or joint discoveries of amateurs and professionals like the Hale-Bopp comet, have been among those that have excited the imagination of the public. Dyson hopes that as in other areas of inquiry, amateurs will bring "new ideas as well as new styles of operation." He argues that in both science and the arts, it is amateurs who "sustain the culture" that allows the professionals to "flourish" (Dyson 1997, 76). There are gains and losses with a new technology; we often are able to have the best of both.

Technology extends our senses. Technology is also much more than an extension. Our intellectual grasp exceeds our biological reach. With telescopes and microscopes we can see farther and smaller. The Hubble telescope circling the earth brings us new discoveries as part of our daily news. We nonastronomers may be thrilled by our heavenly technological pyrotechnic displays on holidays and other celebratory events; astronomers can use their technology to observe far more dazzling cosmic "pyrotechnics."

One of the most recent infrared images "captured by the Hubble Space Telescope is thought to show galaxies as far away as 12 billion light-years, at the edge of the observable universe. The image spans 2 million light-years at its maximum, with just one light-year equaling 6 trillion miles" (Boyle 1998). With the Hubble, an array of new objects in the sky can be explored in greater depth. The Hubble's new camera, NICMOS, "has parted the dark curtain that previously blocked our view of very distant objects and revealed a whole new cast of characters. . . . We are still finding new frontiers" (quoted in Boyle 1998).

As most of us know, light years are a measure of distance, namely the distance traversed by light in a vacuum in a (mean solar) year or approximately 5.88×10 to the 12th power (5.88 trillion) miles. Multiply this by twelve billion and, need we say, this is a distance so great that it is beyond the comprehension of most of us. Equally exciting, it means that since it is the distance that light travels in a year, when we see objects twelve billion light years away, we are seeing them as they existed twelve billion years ago. We are literally looking back in time. In this case, we might be "seeing" the universe as it existed less than a billion years after the time when many believe it was created. Before now, who would have ever believed that this was possible?

Early in 2000 it was announced that the "extreme universe is about to flash into view as never before. Earth's atmosphere blocks X-rays from reaching the ground (even though they travel a short distance through air in doctors' offices), so they must be collected by instruments flying aboard satellites in space" (Glanz 2000).

The exploration of this extreme universe was made possible by the "launching of three new X-ray observatories" in the last half of 1999. What we already know about this "extreme universe" (i.e., the "deep-space phenomena that emit X-rays") is exciting and what we will learn will be fueling our imagination even further (Glanz 2000). For many of us, it is for the present time, beyond our imagination, that the X-ray satellite, Chandra, may be picking up evidence of events—in a very real and important sense "seeing" them—of fourteen billion years ago "before there was light" (Wilford 2000; see also Weisskopf et al. 2000; Hasinger 2000; Canizares 2000). There is the XMM (X-ray Multi-Mirror), the European Chandralike X-ray observatory, also called the Newton. "Electromagnetic waves that are thousands of times more energetic than the visible and ultraviolet light given off by lazy nebulas and placidly burning stars, and invisible to the naked eye, X-rays stream from gigantic explosions, matter smashing together at nearly the speed of light, and gases so hot that they cannot be detected with ordinary telescopes" (Glanz 2000).

The Next Generation Space Telescope is "expected to be launched a million miles into space in 2007 and be able to detect infrared emissions from objects 400 times fainter than those being studied" today. It will have a mirror "nearly three times larger than Hubble's" (Wilford 1998).

New discoveries are being made using a microwave telescope named BOOMERANG (an acronym for Balloon Observations of Millimetric Extragalactic Radiation and Geophysics) installed in a high-altitude balloon raised over Antarctica (De Bernardis et al. 2000; Seife 2000; Hu 2000). Once again there were measurements of the tiny variations in the background radiation that are assumed to be remnants of the big bang. BOOMERANG was very quickly followed by Maxima 1 and 2 (with 3 to follow shortly), which have given even more details about the flatness of the universe and more information about its very early history. With BOOMERANG and Maxima, astronomers are detecting the afterglow of creation and the harmonic patterns of the soundwaves that it created.

Epilogue to the Epilogue

The integrating thesis of all of my life's work including this book, separately or taken together, is that modern science and technology give us the opportunity, as never before, to create a better world for all. A better world for all is a better world for each of us, not only in the ethical and moral sense of concern for our fellow human inhabitants, but also in the sense of making it a safer world. My major concern now and in all my previous work is not whether we can do it but whether we have the will and understanding to do it. Yes, we are our brothers' and sisters' keepers, all of them, because for the first time in human history we have the means to do so. Tragically, those shouting loudest about saving the earth and defending its poorest inhabitants are those most actively working to keep us from developing the science and technology to do so. Never in my adult life have the science and the scientist been so overwhelmingly in support of a technology as is the case with biotechnology in agriculture and a range of other technologies in food production and human health. Never has the opposition been so organized and the media and public so effectively misled on these issues. Clearly much remains to be done in public science education.

Being an optimist, I write books and articles with the uncompromising and undiminished faith that the light of reason will shine through the darkness of even the most organized ignorance, and that science, technology, and other human knowledge and understanding will show us the way to that future that we all desire and that the least privileged amongst us desperately need. To that quiet voice of despair that whispers the question "Is there hope for humanity?" we should answer with a resounding YES!

References

Abbott, Alison. 2002. Cancer Research: On the Offensive. *Nature* 416(6880):470–74, 4 April.

Ackerman, Todd. 2002. "Major Step" in Cancer Fight: Drug Said to Shrink Lung Tumors in 10% of Patients. *Houston Chronicle,* 18 May.

Agarwal, Anil, and Sunita Narain. 1991. Chipko People Driven to Jungle Kato Stir. *Economic Times* (India), 31 March.

Agarwal, Radha Raman. 1965. *Soil Fertility in India.* Bombay: Asia Publishing House.

Aiello, Leslie C., and Peter Wheeler. 1995. The Expensive-tissue Hypothesis: The Brain and the Digestive System in Human and Primate Evolution. *Current Anthropology* 36:199–221.

Allen, Arthur. 2002. Bucking the Herd: Parents Who Refuse Vaccination for Their Children May Be Putting Entire Communities at Risk. *The Atlantic* 290(2):40, 42, September.

Aly, Gotz, Peter Chroust, and Christian Pross. 1994. *Cleansing the Fatherland: Nazi Medicine and Racial Hygiene* (translated by Belinda Cooper; foreword by Michael H. Kater). Baltimore: The Johns Hopkins University Press.

Ames, Bruce N., Margie Profet, and Lois Swirsky Gold. 1990a. Dietary Pesticides (99.9% All Natural). *Proceedings of the National Academy of Sciences USA* 87:7777–81.

_____. 1990b. Nature's Chemicals and Synthetic Chemicals: Comparative Toxicology. *Proceedings of the National Academy of Sciences USA* 87:7782–86.

_____. 1983. Dietary Carcinogens and Anticarcinogens: Oxygen Radicals and Degenerative Diseases. *Science* 221(4617)1256–64, 23 September.

Ammann, Klaus. 2002. Mexican Maize Madness, A Summary From ABC's Anna Salleh. Klaus Ammann's Listserv, 10 July.

Anbar, A. D., and A. H. Knoll. 2002. Proterozoic Ocean Chemistry and Evolution: A Bioinorganic Bridge? *Science* 297(5584):1137–42, 16 August.

Angier, Natalie. 2002. Cooking, and How It Slew the Beast Within. *New York Times*, 28 May.

Anker, Peder. 2001. *Imperial Ecology: Environmental Order in the British Empire, 1895–1945*. Cambridge, MA.: Harvard University Press.

Antoniou, Michael. 2000. A Geneticist View of the Dangers of GM. *Ethical Consumer Magazine*, June/July.

AP (Associated Press). 2002. Conference Hears How 3 Million Children Die from Bad Environment. Bangkok, Thailand: Associated Press online, 3 March.

Arluke, Arnold, and Boria Sax. 1992. Understanding Nazi Animal Protection and the Holocaust. *Anthrozoos* 5(1):6–31.

Armelagos, George J., Dennis P. Van Gerven, Debra L. Martin, and Rebecca Huss-Ashmore. 1984. Effects of Nutritional Change on the Skeletal Biology of Northeast African (Sudanese Nubian) Populations. In *From Hunters to Farmers: The Causes and Consequences of Food Production in Africa*, edited by J. Desmond Clark and Steven A. Brandt, pp. 132–47. Berkeley: University of California Press.

Asano, Shin-Ichiro, Yuki Nukumizu, Hisanori Bando, Toshihiko Iizuka, and Takashi Yamamoto. 1997. Cloning of Novel Enterotoxin Genes from *Bacillus cereus* and *Bacillus thuringiensis*. *Applied and Environmental Microbiology* 63(3):1054–57, March.

Asimov Isaac. 1962. *The Search for the Elements*. New York: Basic Books.

Atkins, John F., and Ray Gesteland. 2002. The 22nd Amino Acid. *Science* 296(5572):1409–10, 24 May.

Avery, Alex. 2000. Vandana Shiva Antoinette: Let Them Eat Weeds! *Global Food Quarterly* (30):6, Spring.

_____. 2002a. Nature's Toxic Tools: The Organic Myth of Pesticide-Free Farming. *Center for Global Food Issues*, 23 October.

_____. 2002b. Warning—Organic Foods Contain Higher Levels Of Chemical Dangerous To Infants. *AgBioView* online, 16 March.

Avery, Alex, and Dennis Avery. 2002. The Soil Association Finally Admits, In a New Report, That the "Perception That Organic Food Is Better for You" Appears to Have Been Largely Based on Intuition Rather than Conclusive Evidence." *Life Sciences Network* online, 12 February.

Avery, Dennis. T. 2002. The Most Sustainable Farming in History Gives The World Its Finest Food Choices: A Response to The Johns Hopkins University Authors, Center for Global Food Issues, Hudson Institute, April.

Aziz, Philippe. 1976. *Doctors of Death* (tranlated from the French by Edouard Bizub and Philip Haentzler under the guidance of Linda Marie de Turenne). Geneva: Ferni.

Bada, Jeffrey L., and Antonio Lazcano. 2002. Origin of Life: Some Like It Hot, But Not the First Biomolecules. *Science* 296(5575):1982–1983, 14 June.

Baeyer, Hans Christian Von. 2000. The Lotus Effect. *The Sciences* 40(1):12–15, January/February.

Bailey, Ronald. 2002. From Real Dangers to Phantom Risks: A Historical Perspective on Environmental Threats to Children's Health. In *Are Children More Vulnerable to Environmental Chemicals: Scientific and Regulatory Issues in Perspective,* ed. by Ashlee Dunston and Daland R. Juberg, pp. 112–41. New York: American Council on Science and Health.

Baker, Brian, Charles Benbrook, Edward Groth, and Karen Lutz Benbrook. 2002. Pesticide Residues in Conventional, Integrated Pest Management (IPM)-grown and Organic Foods: Insights from Three US Data Sets. *Food Additives and Contaminants* 19(5):427–46, May.

Balfour, Eve (Lady Evelyn Barbara). 1948. *The Living Soil: Evidence of the Importance to Human Health of Soil Vitality, with Special Reference to National Planning.* New York: Devin-Adair Co.

————. 1976. *The Living Soil and the Haughley Experiment.* New York: Universe Books.

Bandyopadhyay, Jayanta. 1999. Chipko Movement: Of Floated Myths and Flouted Realities. "Mountain People, Forests, and Trees," Mountain Forum's on-line library, http://www.mtnforum.org/resources/library/bandj99a.htm.

Bankier, David. 1994. On Modernization and the Rationality of Extermination. *Yad Vashem Studies* 24():109–29.

Bari, M. L., E. Nazuka, Y. Sabina, S. Todoriki, and K. Isshikla. 2003. Chemical and Irradiation Treatments for Killing *Escherichia coli* O157:H7 on Alfalfa, Radish, and Mung Bean Seeds. *Journal of Food Protection* 66(5):767, 774, May.

Barrett, Stephen. 2001. *Homeopathy: The Ultimate Fake.* Loma Linda, CA.: National Council Against Health Fraud, online.

Bates, Marston. 1967. *Gluttons and Libertines: Human Problems of Being Natural.* New York: Vantage Books.

Bauman, Zygmunt. 1989. *Modernity and the Holocaust.* Ithaca: Cornell University Press.

Baxter, Gwen J., Allan B. Graham, James R. Lawrence, David Wiles, and John R. Paterson. 2001. Salicylic Acid in Soups Prepared from Organically and Non-organically Grown Vegetables. *European Journal of Nutrition* 40(6):289–92.

Beckner, Morton. 1967. Vitalism. In Paul Edwards, ed., *The Encyclopedia of Philosophy*, pp. 253–56. New York: Macmillan Publishing Company, Inc. & The Free Press, Volume Eight.

Benarde, Melvin A. 2002. *You've Been Had!: How the Media and Environmentalists Turned America into a Nation of Hypochondriacs.* New Brunswick, NJ: Rutgers University Press.

Bennetzen, Jeffrey. 2002. The Rice Genome: Opening the Door to Comparative Plant Biology. *Science* 296(5565):60–63, 5 April.

Bentley, D. D., K. F. Chater, A.-M. Cerdeno-Tarraga, G. L. Challis, N. R. Thomson, K. D. James, D. E. Harris, M. A. Quail, H. Kieser, D. Harper, A. Bateman, S. Brown, G. Chandra, C. W. Chen, M. Collins, A. Cronin, A. Fraser, A. Goble, J. Hidalgo, T. Hornsby, S. Howarth, C.-H. Huang, T. Kieser, L. Larke, L. Murphy, K. Oliver, S. O'Neil, E. Rabbinowitsch, M.-A. Rajandream, K. Rutherford, S. Rutter, K. Seeger, D. Saunders, S. Sharp, R. Squares, S. Squares, K. Taylor, T. Warren, A. Wietzorrek, J. Woodward, B. G. Barrell, J. Parkhill, and D. A. Hopwood. 2002. Complete Genome Sequence of the Model Actinomycete *Streptomyces coelicolor* A3(2). *Nature* 417(6885):141–47, 9 May.

Benveniste, J. 1988. Dr. Jacques Benveniste Replies. *Nature* 334(6180):291, 28 July.

Berben, Paul. 1968. *Dachau 1933–1945: The Official History*. London: The Norfolk Press.

Bernstein, Max P., Jason P. Dworkin, Scott A. Sandford, George W. Cooper, and Louis J. Allamandola. 2002. Racemic Amino Acids from the Ultraviolet Photolysis of Interstellar Ice Analogues. *Nature* 416(6879): 401–403, 28 March.

Bhalla, Surjit S. 2002. *Imagine There's No Country: Poverty, Inequality, and Growth in the Era of Globalization*. Washington, DC: Institute for International Economics.

Biehl, Janet, and Peter Staudenmaier. 2000. *Ecofascism: Lessons From the German Experience*. San Francisco: AK Press.

Bittman, Mark. 1994. Eating Well: A Little Cooking Goes a Long Way to Make the Most of Vegetable Nutrients. *New York Times*, 31 August.

Blaine, Katija, and Douglas Powell. 2002. Communication of Food-related Risks. *AgBioForum: A Journal Devoted to the Economics and Management of Agrobiotechnology* online, Special Issue, Communicating About Agrobiotechnology 4(3&4). http://www.agbioforum.org/.

Bob, Clifford. 2000. Beyond Transparency: Visibility and Fit in the Internationalization of Internal Conflict. In Bernard Finel and Kristin M. Lord, eds., *Power and Conflict in the Age of Transparency*, pp. 287–314. New York: Palgrave/St. Martins Press.

_____. 2001. Marketing Rebellion: Insurgent Groups, International Media, and NGO Support. *International Politics* 38(3):311–34, September.

_____. 2002. Merchants of Morality. *Foreign Policy* 36–45, March/April.

Bock, August. 2001. Invading the Genetic Code. *Science* 292:5516(453–54), 20 April.

Bookchin, Murray. 1995. *Re-enchanting Humanity: A Defense of the Human Spirit Against Antihumanism, Misanthropy, Mysticism, and Primitivism*. London and New York: Cassell.

Boyle, Alan. 1998. Hubble Picture Pushes the Frontier: Infrared Image Shows Faintest Objects Ever Seen. *MSNBC*, 8 October.

Brace, C. Loring. 1996. Modern Human Origins and the Dynamics of Regional Continuity. In Takeru Akazawa and Emöke Szathmary, eds., *Prehistoric Mongoloid Dispersals,* pp. 81–112. New York: Oxford University Press.

_____. 1999. Comment on: The Raw and the Stolen: Cooking and the Ecology of Human Origins. *Current Anthropology* 40(1):577–79, December.

Bradley, David. 2002. The Genome Chose Its Alphabet With Care. *Science* 297(5588):1789–91, 13 September.

Bramwell, Anna. 1984. Was This Man "Father of the Greens"? *History Today,* 7–13, September.

_____. 1989. *Ecology in the 20th Century: A History.* New Haven: Yale University Press.

Brasher, Philip. 2002. USDA Lays Egg By Requiring Chickens To Go Outdoors, Farmers Say. *Des Moines Register, Associated Press,* 3 October.

Brock, Thomas D. 1988, *Robert Koch: A Life in Medicine and Bacteriology.* Madison, WI: Berlin; New York: Science Tech Publishers: Springer-Verlag.

Bronowski, Jacob. 1965. *Science and Human Values.* New York: Harper & Row.

Brumbley, Jean. 2002. Dangers of the Salicylates in Organic Food. *AgBioView* online, 20 March.

Budiansky, Stephen. 1995. *Nature's Keepers: The New Science of Nature Management.* New York: Free Press.

_____. 2002a. The Future of Life. *Prospect* Issue 73:30–35, April.

_____. 2002b. How Affluence Could Be Good for the Environment. *Nature* 416(6881):581, 11 April.

Burcher, Sam. 2002. Homeopathy Enters the Mainstream. *ISIS* (Institute of Science in Society: Science, Society, Sustainability), http://www.i-sis.org.uk/water2.php, 30 May.

Burleigh, Michael. 1996. Medicine and Books: A Review of: *Reenchanted Science: Holism in German Culture from Wilhelm II to Hitler* by Anne Harrington, in *BMJ (British Medical Journal)* 313(7070):1494, 7 December.

_____. 2002. The Legacy of Nazi Medicine in Context. In *Medicine and Medical Ethics in Nazi Germany: Origins, Practices, Legacies,* edited by Francis R. Nicosia and Jonathan Huener, pp. 112–27. New York: Berghahn Books.

Byrne, Celia. 2000. Risk Factors: Breast Cancer, Bethesda, Maryland: Environmental Epidemiology Branch, Division of Cancer Etiology, National Cancer Institute, National Institute of Health.

Campbell, Mary Anne, and Randall K. Campbell-Wright. 1995. Toward a Feminist Algebra. In *Teaching the Majority: Breaking the Gender Barriers in Science, Mathematics and Engineering,* edited by Sue V. Rosser, pp. 127–44. New York: Teachers College Press.

Canizares, Claude R. 2000. X-ray Visionaries. *The Sciences* 40(3):26–31, May/June.

Cantrell, Ronald P., and Timothy G. Reeves. 2002. The Rice Genome: The Cereal of the World's Poor Takes Center Stage. *Science* 296(5565):53, 5 April.

Carlson, David S., and Dennis P. Van Gerven. 1977. Masticatory Function and Post-Pleistocene Evolution in Nubia. *American Journal of Physical Anthropology* 46:495–506.

Cartmill, Matt. 1993. *A View to a Death in the Morning: Hunting and Nature Through History.* Cambridge: Harvard University Press.

Castellani, E. 2002. Reductionism, Emergence, and Effective Field Theories. *Studies in History and Philosophy of Science Part B: Studies in History and Philosophy of Modern Physics* 33(2):251–67, June.

CBU (Crop Biotech Update). CIMMYT Repeats: No GM in Maize Gene Banks. *Crop Biotech Update,* 17 May.

CGHFBC (Collaborative Group on Hormonal Factors in Breast Cancer). 2002. Breast Cancer and Breastfeeding: Collaborative Reanalysis of Individual Data from 47 Epidemiological Studies in 30 Countries, Including 50 302 Women With Breast Cancer and 96 973 Women Without the Disease. *The Lancet* 360(9328), 20 July.

Chase, Alston. 1995. *In a Dark Wood: The Fight Over Forests and the Rising Tyranny of Ecology.* Boston: Houghton Mifflin Co.

Chassy, Bruce M. 2002. Food Safety Evaluation of Crops Produced through Biotechnology. *Journal of the American College of Nutrition* (Special Supplement—*The Future of Food and Nutrition With Biotechnology*) 21(3):166S–173S, June.

Chown, Marcus. 1998. Stars in Their Eyes. *New Scientists* 159(2141), 4 July.

Christou, Paul. 2002. No Credible Scientific Evidence is Presented to Support Claims that Transgenic DNA was Introgressed into Traditional Maize Landraces in Oaxaca, Mexico. *Transgenic Research* 11(1):3–5, February.

CIMMYT (Centro Internacional de Mejoramiento de Maiz y Trigo). 2002. Transgenic Maize in Mexico: Facts and Future Research Needs. *Centro Internacional de Mejoramiento de Maiz y Trigo* (International Maize and Wheat Improvement Center), 8 May.

Clarke, Tom. 2002. Bugs Dress Salad: Harmful Bugs May Lurk Within Leaves. *Nature Science Update* online, 7 January.

CNS. 1998. Environmental Pollution and Degradation Causes 40 Percent of Deaths Worldwide, Cornell Study Finds. Cornell News Service, 30 September.

Coggon, David, and Hazel Inskip. 1994. Current Issues in Cancer: Is There an Epidemic of Cancer? *BMJ* (*British Medical Journal*) 308(6930):705–708, 12 March.

Commoner, Barry. 1968. Failure of the Watson-Crick Theory as a Chemical Explanation of Inheritance. *Nature* 220:334–40, 26 October.

_____. 2002. Unraveling the DNA Myth: The Spurious Foundation of Genetic Engineering. *Harpers Magazine* 304(1821), February.

Conko, Gregory, and C. S. Prakash. 2002. Report of Transgenes in Mexican Corn Called Into Question. *ISB News Report* (Information Systems For Biotechnology):3–5. March.

Cooper, George, Novelle Kimmich, Warren Belisle, Josh Sarinana, Katrina Brabham, and Laurence Garrel. 2001. Carbonaceous Meteorites as a Source of Sugar-related Organic Compounds for the Early Earth. *Nature* 414(6866):879–83, 20/27 December.

Corbey, Raymond, and Wil Roebroeks. 1997. Ancient Minds: A Review Essay of *The Prehistory of the Mind: A Search for the Origins of Art, Religion and Science* (London: Thames and Hudson, 1996) by Steven Mithen and *Human Evolution, Language, and Mind: A Psychological and Archaeological Inquiry* (Cambridge: Cambridge University Press, 1996) by William Noble and Iain Davidson. *Current Anthropology* 38(5):917–21, December.

CR. 2002. Organic Its Lower in Pesticides. Honest. *Consumer Reports* 67(8):6, August

Crick, Francis H. C. 1958. On Protein Synthesis. In The Biological Replication of Macromolecules. *Symposia of the Society of Experimental Biology* XII:138–63, Cambridge: Cambridge University Press.

Crick, Sir Francis H. C. 1970. Central Dogma of Molecular Biology. *Nature* 227(1198):561–63, 8 August.

CSIRI. 2002. Seafood Retains Healthy Oil After Cooking. *CSRI* (*Commonwealth Scientific and Industrial Research Organization*) Online, 16 May.

CUPR. 2002. Consumers Union Research Team Shows: Organic Foods Really Do Have Less Pesticides. Consumers Union—Press Release, 8 May.

Dabis, F., J Orne-Gliemann, F. Perez, V. Leroy, M. L. Newell, A. Coutsoudis, and H. Coovadia (Working Group on Women and Child Health). 2002. Education and Debate: Improving Child Health: The Role of Research. *BMJ* (*British Medical Journal*) 324(7351):1444–47, 15 June.

Dale, Philip J., Belinda Clarke, and Eliana M.G. Fontes. 2002. Research Review: Environmental Impact of GM Crops Potential for the Environmental Impact of Transgenic Crops. *Nature Biotechnology* 20(6):567–74, June.

Davenas, E., F. Beauvais, J. Arnara, M. Oberbaum, B. Robinzon, A. Miadonna, A. Tedeschi, B. Pomeranz, P. Fortner, P. Belon, J. Sainte-Laudy, B. Poitevin, and Jacques Benveniste. 1988. Human Basophil Degranulation Triggered by Very Dilute Antiserum Against IgE. *Nature* 333(6176):816–18, 30 June.

Davidson, Sir Stanley, R. Passmore, J. F. Brock, and A. S. Truswell. 1979. *Human Nutrition and Dietetics* (seventh edition). Edinburgh: Churchill and Livingston.

Day, Stephen. 2001. Ag Researchers In The United States Prepare To Harvest The World's First Hypoallergenic Wheat. *The Guardian* (London), 24 May.

Deaton, Angus. 2002. Is World Poverty Falling? *Finance & Development* 39(2):4–7, June.

De Bernardis, P., P. A. R. Ade, J. J. Bock, J. R. Bond, J. Borrill, A. Boscaleri, K. Coble, B. P. Crill, G. De Gasperis, P. C. Farese, P. G. Ferreira, K. Ganga, M. Giacometti, E. Hivon, V. V. Hristov, A. Iacoangeli, A. H. Jaffe, A. E. Lange, L. Martinis, S. Masi, P. V. Mason, P. D. Mauskopf, A. Melchiorri, L. Miglio, T. Montooy, C. B. Netterfield, E. Pascale, F. Piacentini, D. Pogosyan, S. Pruney, S. Rao, G. Romeo, J. E. Ruhl, F. Scaramuzzi, D. Sforna, and N. Vittorio. 2000. A Flat Universe from High-Resolution Maps of the Cosmic Microwave Background Radiation. *Nature* 404(6781):955–959, 27 April.

de Chadarevian, Soraya. 2002. *Life Molecular Biology After World War II*. New York: Cambridge University Press.

De Costa, Caroline. 2002. St Anthony's Fire and Living Ligatures: A Short History of Ergometrine. *The Lancet* 359(9319):1768–70, 18 May.

DeGregori, Thomas R. 1974. *Technology and Economic Change: Essays and Inquiries*. Canada, McLoughlin Associates, Ltd., P.O. Box 1288, Comox, British Columbia, Canada, Monograph Series: *Notes and Papers in Development*, No. 9.

_____. 1985. *A Theory of Technology: Continuity and Change in Human Development*. Ames: Iowa State University Press.

_____. 1987. Technology in Our Changing World. In *Symposium on Science and Technology for Development: Prospects Entering the 21st Century*. Washington, DC: National Academy of Sciences and U.S. AID, June.

_____. 2001. *Agriculture and Modern Technology: A Defense*. Ames: Iowa State University Press.

_____. 2002a. Dangers of the Salicylates in Organic Food. *AgBioView* on-line, 20 March.

_____. 2002b. *The Environment, Our Natural Resources, and Modern Technology*. Ames: Iowa State Press, A Blackwell Scientific Publisher.

_____. 2002c. Warning—Organic Foods Contain Higher Levels Of Chemical Dangerous To Infants. *AgBioView* online, 16 March.

_____. 2003. DNA and Reductionist Science, American Council on Science and Health: HealthFacts&Fears, 25 March.

Deichmann, Ute. 2000. An Unholy Alliance: The Nazis Showed That "Politically Responsible" Science Risks Losing Its Soul. _Nature_ 405(6788):739, 15 June.

Demmig-Adams, Barbara, and William W. Adams III. 2002. Antioxidants in Photosynthesis and Human Nutrition. _Science_ 298(5601):2149–53, 13 December.

Denier van der Gon, H. A. C., M. J. Kropff, N. van Breemen, R. Wassmann, R. S. Lantin, E. Aduna, T. M. Corton, and H. H. van Laar. 2002. Optimizing Grain Yields Reduces CH4 Emissions From Rice Paddy Fields. _PNAS_ (Proceedings of the National Academy of Sciences) online 20 August, 10.1073/pnas.192276599.

Derbyshire, David. 2002. Adviser Accuses BBC of Being Anti-GM in "Ridiculous" Thriller. _Daily Telegraph_ (London), 31 May.

Devi, Savitri. 1991. _Impeachment of Man._ Costa Mesa, CA.: Noontide Press.

Dewanto, Veronica, Xianzhong Wu, and Rui Hai Liu. 2002. Processed Sweet Corn Has Higher Antioxidant Activity. _Journal of Agricultural and Food Chemistry_ 50(17):4959–64, 14 August.

Dewanto, Veronica, Xianzhong Wu, Kafui K. Adom, and Rui Hai Liu. 2002. Thermal Processing Enhances the Nutritional Value of Tomatoes by Increasing Total Antioxidant Activity. _Journal of Agricultural and Food Chemistry_ 50(10):3010–14, 17 April.

Dewey, John. 1929. _The Quest for Certainty: A Study of the Relation of Knowledge and Action._ 1980 reprint, New York: Capricorn Books, G.P. Putnam & Sons.

Dickstein, Morris. 2003. Literary Theory and Historical Understanding, The Chronicle Review. _Chronicle of Higher Education_ 49(37):B7, 23 May.

Diner, Dan. 1994. Rationalization and Method: Critique of a New Approach in Understanding the Holocaust. _Yad Vashem Studies_ 24:87.

_____. 2000. _Beyond the Conceivable: Studies on Germany, Nazism, and the Holocaust._ Berkeley and Los Angeles: University of California Press.

Ditt, Renata F., Eugene W. Nester, and Luca Comai. 2001. Plant Biology Plant Gene Expression Response to _Agrobacterium tumefaciens._ _PNAS_ (Proceedings of the National Academy of Sciences) 98(19):10954–59, 11 September.

Dixon, Bernard. 1988. Criticism Builds Over Nature Investigation. _The Scientist_ 2(16):1, 5 September.

Doering, Volker, Henning D. Mootz, Leslie A. Nangle, Tamara L. Hendrickson, Valerie de Crecy-Lagard, Paul Schimmel, and Philippe Marliere. 2001. Enlarging the Amino Acid Set of _Escherichia coli_ by Infiltration of the Valine Coding Pathway. _Science_ 292:5516(501–504), 20 April.

Doll, Richard. 1992. Health and the Environment in the 1990s. *American Journal of Public Health* 82(7):933–41, 1 July.

Doll, Richard, and Richard Peto. 1981. The Causes of Cancer: Quantitative Estimates of Avoidable Risks of Cancer in The United States Today. *Journal of the National Cancer Institute* 66(6):1191–1308, June.

_____. 1992. *The Environmental Movement in Germany: Prophets and Pioneers, 1871–1971.* Bloomington: Indiana University Press.

Dorsey, Michael K. 2002. The New Eugenics. *World Watch: Working For a Sustainable Future* 15(4):21–23, July/August.

Dreifus, Claudia. 2001. A "Dark Remedy" Now Is Generating Light. *New York Times*, 31 July.

Driesch, Hans. 1908. *The Science and Philosophy of the Organism.* London: A. C. Black.

Driesch, Hans, and C. K. Ogden. 1914. *The History and Theory of Vitalism* (Rev. and in part rewritten for the English ed). London: Macmillan and Co.

Dvorak, John C. 1998. Inside track. *PC Magazine* 17(19): 89, 3 November.

Dyson, Freeman J. 1997. *Imagined Worlds.* Cambridge: Harvard University Press.

_____. 1999. Miracles of Rare Device: New Tools, Not New Ideas, Will Usher in Tomorrow's Scientific Marvels. *The Sciences* 39(2), March/April.

Economist. 2001. Lexington: Leon Kass, Philospher-politician. *The Economist* 360(8235):41, 18 August.

Ehrenfeld, David. 1978. *The Arrogance of Humanism.* New York: Oxford University Press.

Einsle, Oliver, F. Akif Tezcan, Susana L. A. Andrade, Benedikt Schmid, Mika Yoshidar, James B. Howard, and Douglas C. Rees. 2002. Nitrogenase MoFe-Protein at 1.16 _ Resolution: A Central Ligand in the FeMo-Cofactor. *Science* 297(5587):1696–1700, 6 September.

Evans, Michael, and Iain Rodger. 2000. *Healing for Body, Soul and Spirit: An Introduction to Anthroposophical Medicine.* Edinburg: Floris Books.

Evenson, R. E., and D. Gollin. 2003. Assessing the Impact of the Green Revolution, 1960 to 2000. *Science* 300(5620): 758–62, 2 May.

FAO. 2002. *World Agriculture: Towards 2015/2030.* Rome: Food and Agriculture Organization of the United Nations, FAO Press.

_____. 2003. *Unlocking the Water Potential of Agriculture.* Rome: Food and Agriculture Organization of the United Nations.

Farmelo, Graham, editor. 2002 *It Must Be Beautiful: Great Equations of Modern Science.* London: Granta.

Feder, Gene, and Tessa Katz. 2002. Randomised Controlled Trials for Homoeopathy. *BMJ* (*British Medical Journal*) 324(7336):498–99, 2 March.

Feldbaum, Carl B. 2002. Biotechnology: Some History Should Be Repeated. *Science* 295(5557):975, 8 February.

Felsot, Allan S. 2002. Some Corny Ideas About Gene Flow and Biodiversity. *Agrichemical and Environmental: A Monthly Report on Environmental and Pesticide Related Issues,* Issue 193, May.

Ferguson, Kirsty. 1997. Steiner's Philosophy on Compost: The Plot Thickens. *The Independent* (London), 1 November.

Ferre, Juan, and Jeroen Van Rie. 2002. Biochemistry and Genetics of Insect Resistance to *Bacillus thuringiensis*. *Annual Review of Entomology* 47:501–33.

Ferris, Timothy. 1998. Annals of Astronomy: Seeing in the Dark—Why are Amateur Stargazers, Armed with Little More Than Homemade Equipment, Making Some of the Biggest Discoveries in Space? *The New Yorker* 74(23):54–61, 10 August.

————. 2002. *Seeing in the Dark: How Backyard Stargazers Are Probing Deep Space and Guarding Earth From Interplanetary Peril.* New York: Simon & Schuster.

Ferry, Luc. 1995. *The New Ecological Order.* Chicago: University of Chicago Press.

Ferry, Luc, and Alain Renaut. 1990. *Heidegger and Modernity* (translated by Franklin Philip). Chicago: University of Chicago Press.

Fest, Joachim C. 1974. *Hitler* (translated from the German by Richard and Clara Winston). New York: Harcourt Brace Jovanovich.

Feynman, Richard Phillips. 1964. *Lectures on Physics: Exercises.* Reading, MA: Addison-Wesley Pub. Co.

Fienberg, Howard. 2001. Medicine v Magic: The Homeopathy Scam, Spiked-online.com/articles/00000002D1E2.htm, 9 August.

————. 2002. A New Organic Stew. *TCS* (Tech Central Station) online, 22 April.

Fitzpatrick, Michael. 2002. Put Alternative Medicine Back in Its Box, Spiked-online.com, 26 June.

————. 2003. Immune to the Facts. Spiked-online.com, 27 May.

Fleischman P. 1970. The Chemical Basis of Inheritance. *Nature* 225():30–32, 3 January.

Fletcher, Liz. 2001. New Zealand GMO Debacle Undermines Green Lobby. *Nature Biotechnology* 19(4):292, April.

Foner, Henry. 1993. Wagner, the Nazis, and Animal Rights. *CFAAR Newsletter* 5(1):10–11, Winter/Spring.

Fowler, William A. 1984. The Quest for the Origins of the Elements. *Science* 226(4677), 23 November.

Friedlander, Saul. 2002. True Believers: Greed, Ideology, Power and Lust as Motives for the Holocaust: A Review of Robert S. Wistrich *Hitler and the Holocaust*, New York: Modern Library. In *TLS, The Times Literary Supplement* 5161, 4–6, 1 March.

Friedmann, John, and Haripriya Rangan. 1993. Introduction. In *In Defense of Livelihood: Comparative Studies on Environmental Action,* edited by John Friedmann and Haripriya Rangan, pp. 1–21. West Hartford, CT.: Kumarian Press.

Frink, Charles R., Paul E. Waggoner, and Jesse H. Ausubel. 1999. Perspective Nitrogen Fertilizer: Retrospect and Prospect. *PNAS (Proceedings of the National Academy of Sciences)* 96(4):1175–80, 16 February.

Fruton, Joseph S. 1999. *Proteins, Enzymes, Genes: The Interplay of Chemistry and Biology.* New Haven, CT: Yale University Press.

Fukuyama, Francis. 2002. *Our Posthuman Future: Consequences of the Biotechnology Revolution.* New York: Farrar Straus & Giroux.

Fulmer, Melinda. 2003. Advocates for Farm Laborers Seek a Ban on Hand Weeding: Cal/OSHA Will Decide on a Request to Prohibit the Practice in the State's Commercial Agriculture, But Growers Say It Is Vital to Many Crops. *Los Angeles Times,* 21 April.

Gahan, Linda J., Fred Gould, and David G. Heckel. 2001. Identification of a Gene Associated with Bt Resistance in Heliothis virescens. *Science* 293(5531):857–60, 3 August.

Galibert, Francis, Turlough M. Finan, Sharon R. Long, Alfred Pühler, Pia Abola, Frédéric Ampe, Frédérique Barloy-Hubler, Melanie J. Barnett, Anke Becker, Pierre Boistard, Gordana Bothe, Marc Boutry, Leah Bowser, Jens Buhrmester, Edouard Cadieu, Delphine Capela, Patrick Chain, Alison Cowie, Ronald W. Davis, Stéphane Dréano, Nancy A. Federspiel, Robert F. Fisher, Stéphanie Gloux, Thérèse Godrie, André Goffeau, Brian Golding, Jérôme Gouzy, Mani Gurjal, Ismael Hernandez-Lucas, Andrea Hong, Lucas Huizar, Richard W. Hyman, Ted Jones, Daniel Kahn, Michael L. Kahn, Sue Kalman, David H. Keating, Ernö Kiss, Caridad Komp, Valérie Lelaure, David Masuy, Curtis Palm,6 Melicent C. Peck, Thomas M. Pohl, Daniel Portetelle, Bénédicte Purnelle, Uwe Ramsperger, Raymond Surzycki, Patricia Thébault, Micheline Vandenbol, Frank-J. Vorhölter, Stefan Weidner, Derek H. Wells, Kim Wong, Kuo-Chen Yeh, and Jacques Batut. 2001. The Composite Genome of the Legume Symbiont Sinorhizobium meliloti. *Science* 293(5530):668–72, 27 July.

Gammon, Marilie D., Mary S. Wolff, Alfred I. Neugut, Sybil M. Eng, Susan L. Teitelbaum, Julie A. Britton, Mary Beth Terry, Bruce Levin, Steven D. Stellman, Geoffrey C. Kabat, Maureen Hatch, Ruby Senie, Gertrud Berkowitz, H. Leon Bradlow, Gail Garbowski, Carla Maffeo, Pat Montalvan, Margaret Kemeny, Marc Citron, Freya Schnabel, Allan Schuss, Steven Hajdu, Vincent Vinceguerra, Nancy Niguidula, Karen Ireland, and Regina M. Santella. 2002. Environmental Toxins and Breast Cancer on Long Island. II. Organochlorine Compound Levels in Blood. *Cancer Epidemiology, Biomarkers & Prevention* 11(8):686–97, August.

Gammon, Marilie D., Regina M. Santella, Alfred I. Neugut, Sybil M. Eng, Susan L. Teitelbaum, Andrea Paykin, Bruce Levin, Mary Beth Terry, Tie Lan Young, Lian Wen Wang, Qiao Wang, Julie A. Britton, Mary S. Wolff, Steven D. Stellman, Maureen Hatch, Geoffrey C. Kabat, Ruby Senie, Gail Garbowski, Carla Maffeo, Pat Montalvan, Gertrud Berkowitz, Margaret Kemeny, Marc Citron, Freya Schnabel, Allan Schuss, Steven Hajdu, and Vincent Vinceguerra. 2002. Environmental Toxins and Breast Cancer on Long Island. I. Polycyclic Aromatic Hydrocarbon DNA Adducts. *Cancer Epidemiology, Biomarkers & Prevention* 11(8):677–85, August.

Gamwell, Lynn. 2002. *Exploring the Invisible: Art, Science, and the Spiritual.* Princeton, NJ: Princeton University Press.

_____. 2003. Perceptions of Science: Beyond the Visible—Microscopy, Nature, and Art. *Science* 299(5603):49–50, 3 January.

Garfield, Simon. 2001. *Mauve: How One Man Invented a Color That Changed the World.* New York: Norton.

Garn, Stanley M. 1994. Uses of the Past. *American Journal of Human Biology* 6(1):89–96.

Gasman, Daniel. 1971. *The Scientific Origins of National Socialism: Social Darwinism in Ernst Haeckel and the German Monist League.* London: Macdonald & Co.

Gatehouse, Angharad M. R., Natalie Ferry, and Romaan J. M. Raemaekers. 2002. The Case of the Monarch Butterfly: A Verdict Is Returned. *Trends in Genetics* 18(5), May.

Gay, Peter. 1968. *Weimar Culture: The Outsider As Insider.* New York: Harper & Row.

Geisler, Charles. 2002. Endangered Humans: How Global Land Conservation Efforts Are Creating a Growing Class of Invisible Refugees. *Foreign Policy* 80–81, May/June.

Giovannucci, Edward, Eric B. Rimm, Yan Liu, Meir J. Stampfer, and Walter C. Willett. 2002. A Prospective Study of Tomato Products, Lycopene, and Prostate Cancer Risk. *Journal of the National Cancer Institute* 94(5):391–98, 6 March.

Glanz, James. 2000. "Extreme" Cosmic Images Start Flashing into View. *New York Times*, January 11.

Goering, Herman. 1939. A Broadcast Over German Radio Network Describing the Fight Against Vivisection and Measures Taken To Prohibit It, August 28, 1933. In *The Political Testament of Herman Goering* (translated by H. W. Blood-Ryan), pp. 70–75. London: John Long.

Goff, Stephen A., Darrell Ricke, Tien-Hung Lan, Gernot Presting, Ronglin Wang, Molly Dunn, Jane Glazebrook, Allen Sessions, Paul Oeller, Hemant Varma, David Hadley, Don Hutchison, Chris Martin, Fumiaki

Katagiri, B. Markus Lange, Todd Moughamer, Yu Xia, Paul Budworth, Jingping Zhong, Trini Miguel, Uta Paszkowski, Shiping Zhang, Michelle Colbert, Wei-lin Sun, Lili Chen, Bret Cooper, Sylvia Park, Todd C. Wood, Long Mao, Peter Quail, Rod Wing, Ralph Dean, Yeisoo Yu, Andrey Zharkikh, Richard Shen, Sudhir Sahasrabudhe, Alun Thomas, Rob Cannings, Alexander Gutin, Dmitry Pruss, Julia Reid, Sean Tavtigian, Jeff Mitchell, Glenn Eldredge, Terri Scholl, Rose Mary Miller, Satish Bhatnagar, Nils Adey, Todd Rubano, Nadeem Tusneem, Rosann Robinson, Jane Feldhaus, Teresita Macalma, Arnold Oliphant, and Steven Briggs. 2002. A Draft Sequence of the Rice Genome (*Oryza sativa L.* ssp. *japonica*). *Science* 296(5565):92–100, 5 April.

Goklany, Indur M. 2002. *The Globalization of Human Well-Being.* Washington, DC: Cato Policy Analysis No. 447, 22 August.

Goldner, Loren. 2001. The Nazis and Deconstruction: Jean-Pierre Faye's Demolition of Derrida. *German Politics and Society* (published by the Center for European Studies, Harvard). Reprinted in *Vanguard of Retrogression: "Postmodern" Fictions as Ideology in the Era of Fictitious Capital,* ed. by Loren Goldner, pp. 57–60. New York: Queequeg Publications. http://home.earthlink.net/~lrgoldner/fayedeutsch.html.

Goldstein, Inge F., and Martin Goldstein. 2002. *How Much Risk? A Guide to Understanding Environmental Health Hazards.* New York: Oxford University Press.

Goldstein, Jeffrey. 1979. "On Racism and Anti-Semitism in Occultism and Nazism" in Livia Rothkirchen. *Yad Vashem Studies* XIII, Jerusalem.

Gong, Y. Y., K Cardwell, A. Hounsa, S. Egal, P. C. Turner. A. J. Hall and C. P Wild. 2002. Dietary Aflatoxin Exposure and Impaired Growth in Young Children From Benin and Togo: Cross Sectional Study. *BMJ (British Medical Journal)* 325(7354):20–21, 6 July.

Goodrick-Clarke, Nicholas. 1992. *The Occult Roots of Nazism: Secret Aryan Cults and their Influence on Nazi Ideology: The Ariosophists of Austria and Germany, 1890–1935.* New York: New York University Press.

————. 1998. *Hitler's Priestess: Savitri Devi, the Hindu-Aryan Myth, and Neo-Nazism.* New York: New York University Press.

————. 2002. *Black Sun: Aryan Cults, Esoteric Nazism and the Politics of Identity.* New York: New York University Press.

Gordimer, Nadine. 2002. A New Racism, *World Watch: Working For a Sustainable Future* 15(4):17, July/August.

Gould, Stephen Jay. 1977. *Ontogeny and Phylogeny.* Cambridge, Mass.: Belknap Press of Harvard University Press.

————. 2000. Deconstructing the "Science Wars" by Reconstructing an Old Mold. *Science* 287(5451):25326, 14 January.

Gray, Carl. 2002. A Review of: *Twice Dead: Organ Transplants and the Reinvention of Death* by Margaret Lock (Berkeley and Los Angeles: University of California Press), in *BMJ* (*British Medical Journal*) 324(7350):1401, 8 June.

Griffitts, Joel S., Johanna L. Whitacre, Daniel E. Stevens, and Raffi V. Aroian. 2001. Bt Toxin Resistance from Loss of a Putative Carbohydrate-Modifying Enzyme. *Science* 293(5531):860–64, 3 August.

Groening, Gert, and Joachim Wolschke-Bulmahn. 1987. Politics, Planning and the Protection of Nature: Political Abuse of Early Ecological Ideas in Germany, 1933–45. *Planning Perspectives* 2:127–48.

_____. 1992. Some Notes on the Mania for Native Plants in Germany. *Landscape Journal* 11(2):116–26, Fall.

Gross, Paul R., and Norman Levitt. 1994. *Higher Superstition: The Academic Left and Its Quarrels with Science*. Baltimore: The Johns Hopkins University Press.

Grusak, Michael A. 2002. Enhancing Mineral Content in Plant Food Products. *Journal of the American College of Nutrition* (Special Supplement—*The Future of Food and Nutrition With Biotechnology*) 21(3):178S–183S, June.

Guha, Ramachandra. 1989. Radical American Environmentalism and Wilderness Preservation: A Third World Critique. *Environmental Ethics* 11(1):71–83, Spring.

_____. 1997. The Authoritarian Biologist and the Arrogance of Anti-humanism: Wildlife Conservation in the Third World. *The Ecologist* 27(1):14–20.

_____. 1998. Deep Ecology Revisited. In *The Great New Wilderness Debate,* edited by J. Baird Callicott and Michael P. Nelson, pp. 271–79. Athens: University of Georgia Press.

Haack, Susan. 1998. *Manifesto of a Passionate Moderate: Unfashionable Essays*. Chicago: University of Chicago Press.

Hall, Allan. 2002. Anti-vaccine Town Struck by Measles Epidemic: Homoeopaths Who Reject MMR are Blamed for German Outbreak. *The Times* (London), 6 March.

Hamilton, David B. 1999. *Evolutionary Economics: A Study of Change in Economic Thought*. New Brunswick, NJ: Transaction Publishers.

Hanauske-Abel, Hartmut M. 1996. Not a Slippery Slope or Sudden Subversion: German Medicine and National Socialism in 1933. *BMJ* (*British Medical Journal*) 313(7070):1453–63, 7 December.

Hao, Bing, Weimin Gong, Tsuneo K. Ferguson, Carey M. James, Joseph A. Krzycki, and Michael K. Chan. 2002. A New UAG-Encoded Residue in the Structure of a Methanogen Methyltransferase. *Science* 296(5572):1462–66, 24 May.

Harlander, Susan K. 2002. The Evolution of Modern Agriculture and Its Future with Biotechnology. *Journal of the American College of Nutrition* (Special Supplement—*The Future of Food and Nutrition With Biotechnology*) 21(3):161S–5S, June.

Harrington, Anne. 1996. *Reenchanted Science: Holism in German Culture from Wilhelm II to Hitler*. Princeton, NJ: Princeton University Press.

Harris, Henry. 1995. *The Cells of the Body: A History of Somatic Cell Genetics*. Plainview, NY: CSHL Press (Cold Spring Harbor Laboratory Press).

_____. 1999. *The Birth of the Cell*. New Haven, CT: Yale University Press.

_____. 2002. *Things Come to Life: Spontaneous Generation Revisited*. New York: Oxford University Press.

Harrison, Paul, and Fred Pearce. 2000. *AAAS Atlas of Population & Environment* (Foreword by Peter Raven). Berkeley and Los Angeles: University of California Press.

Hasinger, Gunther. 2000. X-ray Astronomy: Peeking into the Obscured Universe. *Nature* 404(6777):443–45, 30 March.

Hawkes, K., J. F. O'Connell, and N. G. Blurton Jones. 1991. Hunting Income Patterns Among the Hadza: Big Game, Common Goods, Foraging Goals, and the Evolution of the Human Diet. *Philosophical Transactions of the Royal Society of London* 334:243–51.

_____. 1999. Comment on: The Raw and the Stolen: Cooking and the Ecology of Human Origins. *Current Anthropology* 40(1):581–82, December.

Heckman, Daniel S., David M. Geiser, Brooke R. Eidell, Rebecca L. Stauffer, Natalie L. Kardos, and S. Blair Hedges. 2001. Molecular Evidence for the Early Colonization of Land by Fungi and Plants. *Science* 293(5532):1129–33, 10 August.

Heidegger, Martin. 1970. *Existence and Being* (Introduction by Weiner Brock). Chicago: Gateway Editions, Henry Regnery.

_____. 1977a. *Basic Writings: From Being and Time (1927) to The Task of Thinking (1964)*, edited by David Farrell Krell. New York: Harper & Row.

_____. 1977b. Letter on Humanism. In *Basic Writings: From Being and Time (1927) to The Task of Thinking (1964)*, edited by David Farrell Krell, pp. 213–65. New York: Harper & Row.

_____. 1977c. *The Question Concerning Technology, and Other Essays* (translated and with an Introduction by William Lovitt). New York: Garland Publisher.

Hellmich, Richard L., Blair D. Siegfried, Mark K. Sears, Diane E. Stanley-Horn, Michael J. Daniels. 2001. Monarch Larvae Sensitivity to *Bacillus thuringiensis*-purified Proteins and Pollen. *PNAS* (Proceedings of the National Academy of Sciences USA) 98, 11 September.

Henderson, Mark. 2002. BBC Drama "Peddles Ludicrous Lies on GM." *The Times* (London), 31 May.

Herf, Jeffrey. 1984. *Reactionary Modernism: Technology, Culture, and Politics in Weimar and the Third Reich.* New York: Cambridge University Press.

Hermand, Jost. 1997. Rousseau, Goethe, Humbolt: Their Influence on Later Advocates of Nature Gardens. In *Nature and Ideology: Natural Garden Design in the Twentieth Century,* edited by Joachim Wolschke-Bulmahn, pp. 155–86. Washington, DC: Dumbarton Oaks Research Library and Collection.

Hilborn, Elizabeth D., Jonathan H. Mermin, Patricia A. Mshar, James L. Hadler, Andrew Voetsch, Christine Wojtkunski Margaret Swartz, Roger Mshar, Mary-Anne Lambert-Fair, Jeffrey A. Farrar, M. Kathleen Glynn, and Laurence Slutsker. 1999. A Multistate Outbreak of *Escherichia coli* O157:H7 Infections Associated With Consumption of Mesclun Lettuce. *Archives of Internal Medicine* 159(15):1758–64, 9/23 August.

Hines, Pamela J. 2001. The Dynamics of Scientific Controversies. *AgBioForum: A Journal Devoted to the Economics and Management of Agrobiotechnology* online, Special Issue, Communicating About Agrobiotechnology, 4(3&4). http://www.agbioforum.org/.

Hirst S. J., M. A. Hayes, J. Burridge, F. L. Pearce, and J. C. Foreman. 1993. Human Basophil Degranulation Is Not Triggered by Very Dilute Antiserum Against IgE. *Nature* 366(6455):525–27, 9 December.

Hitler, Adolf. 1939. *Mein Kampf* (translated by James Murphy). London & New York: Hurst and Blackett.

Hodgson, John. 2002a. Doubts Linger Over Mexican Corn Analysis. *Nature Biotechnology* 20(1):3–4, January.

_____. 2002b. Maize Uncertainties Create Political Fallout. *Nature Biotechnology* 20(2):106–107, February.

Holmes, Frederic Lawrence. 2001. *Meselson, Stahl, and the Replication of DNA: A History of "The Most Beautiful Experiment in Biology."* New Haven, CT: Yale University Press.

Holt, Jim. 2002. "Intelligent Design Creationism and Its Critics": Supernatural Selection: A Review of *Intelligent Design Creationism And Its Critics: Philosophical, Theological, and Scientific Perspectives,* edited by Robert T. Pennock. Cambridge, Mass.: A Bradford Book/The MIT Press. *New York Times Book Review,* 12–13, 14 April.

Hopkins, Sir Frederick Gowland. 1929. The Earliest History of Vitamin Research, Nobel Lecture, December 11, 1929, The Nobel Prize in Physiology or Medicine.

Horkheimer, Max, and Theodor W. Adorno. 2000. *Dialectic of Enlightenment.* New York: Continuum.

Horton, Larry. 1988. A Look at the Politics of Research with Animals: Regaining Lost Perspective. *The Physiologist* 31(3):41–44.

_____. 1989. The Enduring Animal Issue. *Journal of the National Cancer Institute* 81(10):736–43, 22 May.

Hotchkiss, Rollin D. 1979. The Identification of Nucleic Acids as Genetic Determinants. In P. R. Srinivasan, Joseph S. Fruton, and John T. Edsall, eds., *The Origins of Modern Biochemistry: A Retrospect on Proteins*, pp. 321–42. New York: New York Academy of Sciences, Annals of the New York Academy of Sciences, Volume 325, 31 May.

Howard, Albert, Sir. 1940. *Agricultural Testament*. Oxford: Oxford University Press.

Hu, Wayne. 2000. Ringing in the New Cosmology. *Nature* 404(6781), 27 April.

Huang, Jikun, Carl Pray, and Scott Rozelle. 2002. Enhancing the Crops to Feed the Poor. *Nature* 418(6898):678–84, 8 August.

Hulme, David, and Michael Edwards, eds. 1997. *NGOs, States and Donors: Too Close for Comfort?* New York: St. Martin's Press in association with Save the Children.

Hume, Mick. 2002. Is It King Kong? Is It Godzilla? No, It's a Genetically Modified Editor. *The Times* (London), 3 June.

IFST (Institute of Food Science and Technology). 2002. Organic Food Not Healthier Than Conventional, Institute of Food Science and Technology, 5 November.

Ingham, Elaine. 2003. CV posted at the Soil Food website. http://www.soil-foodweb.com/phpweb/

IPS (Inter Press Service). 2002. IRRI Hybrid Rice Said to Threaten Diversity. Bangkok: Inter Press Service via NewsEdge Corporation, 8 August.

IRRI (International Rice Research Institute). 2001. *Rice Research: The Way Forward: IRRI Annual Report 2000–2001*. Los Banos: International Rice Research Institute 11(2), July.

ITA (Institute of Tropical Agriculture). 2002. Fungus Linked to Stunted Growth in West African Children, Institute of Tropical Agriculture Press Release, 11 July.

Jackson, Trevor. 2002. Future Imperfect. *BMJ* (*British Medical Journal*) 324(7351):1438–62, 15 June.

Jacob, Francois. 1977. Evolution and Tinkering. *Science* 196(4295):1161–66, 10 June.

JAMA. 2002. Theme Issue: Peer Review Congress. *JAMA* (*The Journal of the American Medical Association*) 287(21), 5 June.

Janick, Jules. 2002. *History of Horticulture*. Purdue University online, www.hort.purdue.edu/newcrop/history/default.html.

Jardhari, Vijay. 1996. Letter Dated 01 May 1996 from Vijay Jardhari and other Chipko Activists of Teri Garhwal to the Editor of the Star (Selangor).

Jha, Prabhat, Anne Mills, Kara Hanson, Lilani Kumaranayake, Lesong Conteh, Christoph Kurowski, Son Nam Nguyen, Valeria Oliveira Cruz,

Kent Ranson, Lara Vaz, Shengchao Yu, Oliver Morton, and Jeffrey D. Sachs. 2002. Improving the Health of the Global Poor. *Science* 295(5562):2036–39, 15 March.

Judson, Horace Freeland. 1996. *The Eighth Day of Creation: Makers of the Revolution in Biology* Plainview, NY: CSHL Press (Cold Spring Harbor Laboratory).

Kaplan, Hillard S., and Arthur J. Robson. 2002. The Emergence of Humans: The Coevolution of Intelligence and Longevity with Intergenerational Transfers. *PNAS* (Proceedings of the National Academy of Sciences) 99(15):10221–26, 23 July.

Kaplinsky, Nick, David Braun, Damon Lisch, Angela Hay, Sarah Hake, and Michael Freeling. 2002. Biodiversity (Communications arising): Maize Transgene Results in Mexico Are Artifacts (followed by an editorial note). Nature AOP, Published online: 4 April 200. *Nature* 416(6881):601–602, 11 April.

Kasting, James F., and Janet L. Siefert. 2002. Life and the Evolution of Earth's Atmosphere. *Science* 296(5570):1066–68, 10 May.

Katz, S. H. 1987. Food and Biocultural Evolution: A Model for the Investigation of Modern Nutritional Problems. In F. E. Johnston, ed., *Nutritional Anthropology,* pp. 111–22. New York: Alan R. Liss.

Keeley, Lawrence H. 2001. Giving War a Chance. In Glen E. Rice and Steven A. LeBlanc, eds., *Deadly Landscapes: Case Studies in Prehistoric Southwestern Warfare*, pp. 331–42. Salt Lake City: University of Utah Press.

Keller, Evelyn Fox. 2000. *The Century of the Gene.* Cambridge, MA: Harvard University Press.

————. 2002. *Making Sense of Life: Explaining Biological Development with Models, Metaphors, and Machines.* Cambridge, MA.: Harvard University Press.

Kennedy, Donald. 2002. When Science and Politics Don't Mix. *Science* 296(5574):1765, 7 June.

King, Mary-Claire, and Arno G. Motulsky. 2002. Human Genetics: Mapping Human History. *Science* 298(5602): 2342–43, 20 December.

Kolata, Gina. 2002a. Epidemic That Wasn't. *New York Times*, 29 August.

————. 2002b. Looking for the Link. *New York Times*, 11 August.

Kolisko, Eugen, and Lilly Noha Kolisko. 1946. *Agriculture of Tomorrow.* Bournemouth: Edge, Strond, Glos., Kolisko archive.

Kolisko, Lilly Noha. 1938. *The Moon and the Growth of Plants.* Bournemouth: Kolisko Archive, 62 Frederica Rd., Bournemouth.

Korban, Schuyler S., Sergei F. Krasnyanski, and Dennis E. Buetow. 2002. Foods as Production and Delivery Vehicles for Human Vaccines. *Journal of the American College of Nutrition* (Special Supplement—*The Future of Food and Nutrition With Biotechnology*) 21(3):212S–7S, June.

Lagerkvist, Ulf. 1998. *DNA Pioneers and Their Legacy*. New Haven, CT: Yale University Press.

Lagunoff, David. 2002. Portraits of Science: A Polish, Jewish Scientist in 19th-Century Prussia. *Science* 298(5602):2331, 20 December.

Lane, Nick. 2002. *Oxygen: The Molecule That Made the World*. Oxford: Oxford University Press.

Langmuir, Irving, and Robert N. Hall. 1989. Pathological Science. *Physics Today* 42(10):36–48, October.

LeBlanc, Steven A. 1999. *Prehistoric Warfare in the American Southwest*. Salt Lake City: University of Utah Press.

_____. 2003. Prehistory of Warfare: Humans Have Been at Each Others' Throats Since the Dawn of the Species. *Archaeology* 56(3):18–25, May/June.

LeBlanc, Steven A., and Glen E. Rice. 2001. Southwestern Warfare: The Value of Case Studies. In Glen E. Rice and Steven A. LeBlanc, eds., *Deadly Landscapes: Case Studies in Prehistoric Southwestern Warfare*, pp. 1–18. Salt Lake City: University of Utah Press.

LeBlanc, Steven A., and Katherine E. Register. 2003. *Constant Battles*. New York: St. Martin's Press.

Lee, Mike. 2003. Advocates Renew Fight to Limit Hand Weeding. *The Sacramento Bee*, 30 April.

Lee, Tong Ihn, Nicola J. Rinaldi, Francois Robert, Duncan T. Odom, Ziv Bar-Joseph, Georg K. Gerber, Nancy M. Hannett, Christopher T. Harbison, Craig M. Thompson, Itamar Simon, Julia Zeitlinger, Ezra G. Jennings, Heather L. Murray, D. Benjamin Gordon, Bing Ren, John J. Wyrick, Jean-Bosco Tagne, Thomas L. Volkert, Ernest Fraenkel, David K. Gifford, and Richard A. Young. 2002. Transcriptional Regulatory Networks in *Saccharomyces cerevisiae*. *Science* 298(5594):799–804, 25 October.

Lehmann, Ernst. 1934. *Biologischer Wille: Wege und Ziele Biologischer Arbeit im Neuen Reich*. Munchen: J. F. Lehmann.

Leonard, William R. 2002. Food for Thought: Dietary Change Was a Driving Force in Human Evolution. *Scientific American* 287(6):106–15, December.

Levenson, Thomas. 1994. *Measure for Measure: A Musical History of Science*. New York: Simon and Schuster.

Levine, Judith. 2002. What Human Genetic Modification Means for Women. *World Watch: Working For a Sustainable Future* 15(4):26–29, July/August.

Levitt, Norman. 1999. *Prometheus Bedeviled: Science and the Contradictions of Contemporary Culture*. New Brunswick, NJ: Rutgers University Press.

Lienhard, John H. 1988–1997a. Equilibrium. *The Engines of Our Ingenuity.* KUHF syndicated radio broadcast no. 1261, http://www.uh.edu/engines/epi1261.htm.

_____. 1988–1997b. Life and Instability. *The Engines of Our Ingenuity.* KUHF syndicated radio broadcast no. 700, http://www.uh.edu/engines/epi700.htm.

_____. 1988–1997c. Roller Skates, *The Engines of Our Ingenuity.* KUHF syndicated radio broadcast no. 1055, http://www.uh.edu/engines/epi1055.htm.

Lilla, Mark. 2001. *The Reckless Mind: Intellectuals in Politics.* New York: New York Review Books.

Liu, Qing, Surinder Singh, and Allan Green. 2002. High-Oleic and High-Stearic Cottonseed Oils: Nutritionally Improved Cooking Oils Developed Using Gene Silencing. *Journal of the American College of Nutrition* (Special Supplement—*The Future of Food and Nutrition With Biotechnology*) 21(3):205S–211S, June.

Liua, Longjian, Katsumi Ikedaa, and Yukio Yamoria on Behalf of the WHO-CARDIAC Study Group. 2002. Inverse Relationship Between Urinary Markers of Animal Protein Intake and Blood Pressure in Chinese: Results from the WHO Cardiovascular Diseases and Alimentary Comparison (CARDIAC) Study. *International Journal of Epidemiology* 31(1):227–33, February.

Logan, Robert A. 2001. News' Compartmentalization: Implications for Food Biotechnology Coverage. *AgBioForum: A Journal Devoted to the Economics and Management of Agrobiotechnology* online, Special Issue, Communicating About Agrobiotechnology, 4(3&4). http://www.agbioforum.org/.

Lonnerdal, Bo. 2002. Expression of Human Milk Proteins in Plants. *Journal of the American College of Nutrition* (Special Supplement—*The Future of Food and Nutrition With Biotechnology*) 21(3):218S–221S, June.

Losey, John, Linda S. Rayor, and Maureen E. Carter. 1999. Transgenic Pollen Harms Monarch Larvae. *Nature* 399(6733):214, 20 May.

Lucca, Paola, Richard Hurrell, and Ingo Potrykus. 2002. Fighting Iron Deficiency Anemia with Iron-Rich Rice. *Journal of the American College of Nutrition* (Special Supplement—*The Future of Food and Nutrition With Biotechnology*) 21(3):184S–190S, June.

Mackey, Maureen. 2002. The Application of Biotechnology to Nutrition: An Overview. *Journal of the American College of Nutrition* (Special Supplement—*The Future of Food and Nutrition With Biotechnology*) 21(3):157S–160S, June.

Maddox, Brenda. 2002. *Rosalind Franklin: The Dark Lady of DNA.* London: HarperCollins.

_____. 2003. The Double Helix and the "Wronged Heroine." *Nature* 421(6921):407–408, 23 January.

Maddox, John Royden. 1988a. Waves Caused by Extreme Dilution. *Nature* 335(6193):760–63, 27 October.

_____. 1988b. When to Believe the Unbelievable. *Nature* 333(6176):787, 30 June.

Maddox, John Royden, James Randi, and Walter W., Stewart. 1988. High-dilution Experiments a Delusion. *Nature* 334(6180):287–90, 28 July.

Mader, Paul, Andreas Fliebach, David Dubois, Lucie Gunst, Padruot Fried, and Urs Niggli. 2002. Soil Fertility and Biodiversity in Organic Farming. *Science* 296(5573):1694–97, 31 May.

Margalit, Avishai, and Ian Buruma. 2002. Occidentalism. *New York Review of Books* 49(1):4–7, 17 January.

Marks, Leonie, and Nicholas Kalaitzandonakes. 2002. Mass Media Communications About Agrobiotechnology. *AgBioForum: A Journal Devoted to the Economics and Management of Agrobiotechnology* on-line, Special Issue, Communicating About Agrobiotechnology, 4(3&4). http://www.agbioforum.org/.

Martinez-Soriano, Juan Pablo Ricardo, A. M. Bailey, J. Lara-Reyna, and D. S. Leal-Klevezas. 2002. Transgenes in Mexican Maize. *Nature Biotechnology* 20(1):19, January.

Martinez-Soriano, Juan Pablo Ricardo, and D. S. Leal-Klevezas. 2000. Transgenic Maize in Mexico: No Need for Concern. *Science* 287(5457):1399, 25 February.

Matossian, Mary Kilbourne. 1989. *Poisons of the Past: Molds, Epidemics, and History*. New Haven: Yale University Press.

Mattisson, Joel L. 2000. Do Pesticides Reduce Our Total Exposure To Food Borne Toxicants? *NeuroToxicology* 21(1&2):195–202, February/April.

Mayer, Jean. 1989. National and International Issues in Food Policy. Lowell Lecture, Harvard University, 15 May.

Mayor, Susan. 2002. Decoding of Soil Bacterium Genome Points Way to New Antibiotics. *BMJ (British Medical Journal)* 324(7347):1176, 11 May.

Mayr, Ernst. 1982. *The Growth of Biological Thought: Diversity, Evolution, and Inheritance*. Cambridge, MA: Belknap Press

_____. 2001. *What Evolution Is*. New York: Basic Books.

McCollum, E. V. 1957. *History of Nutrition: The Sequence of Ideas in Nutrition Investigations*. Boston: Houghton Mifflin.

McGrayne, Sharon Bertsch. 2001. *Prometheans in the Lab: Chemistry and the Making of the Modern World*. New York: McGraw-Hill.

McGrew, W. C. 1999. Comment on: The Raw and the Stolen: Cooking and the Ecology of Human Origins. *Current Anthropology* 40(1):582–83, December.

McKibben, Bill. 2002. Unlikely Allies Against Cloning. *New York Times*, 27 March.

McNeil, Donald G. 2002. When Parents Say No to Child Vaccinations. *New York Times*, 30 November 30.

Mehta, Roshni A., Tatiana Cassol, Ning Li, Nasreen Ali, Avtar K. Handa, and Autar K. Mattoo. 2002. Engineered Polyamine Accumulation in Tomato Enhances Phytonutrient Content, Juice Quality, and Vine Life. *Nature Biotechnology* 20(6):613–18, June.

Mencken, H. L. 1930. *Treatise on the Gods*. New York: A. A. Knopf.

Mertens, Gilbert. 2001. Comment on the Article "When Our Health Is At Risk, Why Be So Mean?", Written by the Prince of Wales and Published by the Times of London in Its December 30, 2000, Issue. *The Times* (London), January.

Metz, Mathew, and Johannes Futterer. 2002. Biodiversity (Communications arising): Suspect Evidence of Transgenic Contamination (followed by an editorial note), Published online: 4 April 2003. *Nature* 416(6881): 600–601, 11 April.

Milo, R., S. Shen-Orr, S. Itzkovitz, N. Kashtan, D. Chklovskii, and U. Alon. 2002. Network Motifs: Simple Building Blocks of Complex Networks. *Science* 298:(5594):824–27, 25 October.

Milton, Katherine. 1999. Comment on: The Raw and the Stolen: Cooking and the Ecology of Human Origins. *Current Anthropology* 40(1):583–84, December.

Mokyr, Joel. 2002. *The Gifts of Athena: Historical Origins of the Knowledge Economy*. Princeton, N.J.: Princeton University Press.

Moore, David. 2001. *Slayers, Saviors, Servants, and Sex: An Expose of Kingdom Fungi*. New York: Springer.

Morgan, Jason Phipps. 2002. When the Earth Moved: How Earth Scientists Developed the Theory of Plate Tectonics: A Review of *Plate Tectonics: An Insider's History of the Modern Theory of the Earth*, edited by Naomi Oreskes. *Nature* 417(6888):487–88, 30 May.

Mullins, Raymond, and Robert Heddle. 2002. Adverse Reactions Associated with Echinacea: The Australian Experience. *Annals of Allergy, Asthma, & Immunology* 88(1):42–51, January.

Nanda, Meera. 1996. The Science Wars in India. *Dissent* 44(1), Winter.

_____. 1997. "History Is What Hurts": A Materialist Feminist Perspective on the Green Revolution and Its Ecofeminist Critics. In *Materialist Feminism: A Reader in Class, Difference, and Women's Lives,* edited by Rosemary Hennessy and Chrys Ingraham, pp. 364–94. New York: Routledge.

_____. 1998. The Episteme Charity of the Social Constructivist Critics of Science and Why the Third World Should Refuse the Offer. In *A House Built on Sand: Exposing Postmodernist Myths About Science,* edited by Noretta Koertge, pp. 286–311. New York: Oxford University Press.

_____. 2000. Dharma and the Bomb: Post-Modern Critiques of Science and the Rise Reactionary Modernism in India. Paper read at the American Sociological Association, August.

_____. 2001. We Are All Hybrids Now: The Dangerous Epistemology of Post-Colonial Populism. *The Journal of Peasant Studies* 28(2): 162–87.

_____. 2002. *Breaking the Spell of Dharma: A Case for Indian Enlightenment*. Delhi: Three Essays Press.

_____. 2003. *Prophets Facing Backwards: Postmodern Critiques of Science and Hindu Nationalism in India*. New Brunswick, NJ.: Rutgers University Press (forthcoming).

Nature. 2002. Correspondence. *Nature* 417(6892):897–98, 27 June.

Navarro-Gonzalez, Rafael, Christopher P. McKay, and Delphine Nna Mvondo. 2001. A Possible Nitrogen Crisis for Archaean Life Due to Reduced Nitrogen Fixation By Lightning. *Nature* 412: 6842(61–64), 5 July.

NCI (National Cancer Institute). 2002. Long Island Breast Cancer Study Project. Washington D.C.: National Cancer Institute, National Institutes of Health Press Release, 6 August.

Nester, Eugene W., Linda S. Thomashow, Matthew Metz, and Milton Gordon. 2002. *100 Years of* Bacillus thuringiensis*: A Critical Scientific Assessment: A Report from the American Academy of Microbiology*. American Academy of Microbiology of the American Society of Microbiology.

Nietzsche, Friedrich Wilhelm. 1998. *Twilight Of The Idols, Or, How To Philosophize With A Hammer* (translation by Duncan Lodge). New York: Oxford University Press.

NNF (Nutrition News Focus). 2002. Cooked News About Tomatoes. *Nutrition News Focus*, 8 May.

North, Richard D. 1996. A Piece Defending Oil Giant Shell's Role in the Nigerian Delta. *The Independent* (London), 10 November.

NYT (New York Times Editorial). 2002. Breast Cancer Mythology on Long Island. *New York Times*, 31 August.

Oakley, Aaron. 2000. Hating Modern Agriculture. *The New Australian*, No. 151, 10–16 April.

Oeppen, Jim, and James W. Vaupel. 2002. Enhanced: Broken Limits to Life Expectancy. *Science* 296(5570):1029–31, 10 May.

O'Hara, Kathleen. 2000. The Stolen Harvest. *The New Australian*, No. 151, 10–16 April.

Olby, Robert C. 1994. *The Path to The Double Helix: The Discovery of DNA*. New York: Dover Publications.

Olsen, Jonathan. 1999. *Nature and Nationalism: Right-Wing Ecology and the Politics of Identity in Contemporary Germany*. New York: St. Martin's Press.

Olson, Walter. 1999. Benighted Elite: Postmodernist Critics of Science Get Their Comeuppance, *Reason* online, June.

Oltvai, Zoltan N., and Albert-Laszlo Barabasi. 2002. Life's Complexity Pyramid. *Science*, 298:(5594) 763–64, 25 October.

O'Neill, Brendan. 2001. Intensive Farming Debates, 29 March. http://www.Spikedonline.com/Articles/000000005542.htm.

Orenstein, Peggy. 2002. Totally Uncooked. *New York Times Magazine*, 2 September.

Overbye, Dennis. 2002. The Most Seductive Equation in Science: Beauty Equals Truth. *New York Times*, 26 March.

Paarlberg, Robert L. 2002. African Famine, Made in Europe. *Wall Street Journal*, 23 August.

Paine, Lynn Sharpe. 1999. Royal Dutch Shell in Transition. Boston: Harvard Business School Publishing Company (A & B), HBS Case Numbers: N9-300-039 & N9-300-040.

Palevitz, Barry A. 2002a. Corn Goes Pop, Then Kaboom: Nature Regrets Publishing a Paper on Transgene Contamination in Mexico. *The Scientist* 16[9]:18, 29 April.

_____. 2002b. Toxicologists Label GM Foods Safe: Draft Report Is Subject to Membership Approval. *The Scientist* 16(8):22, 15 April.

Palfreman, Jon. 2001. Sending Messages Nobody Wants to Hear: A Primer in Risk Communication. *AgBioForum: A Journal Devoted to the Economics and Management of Agrobiotechnology* online, Special Issue, Communicating About Agrobiotechnology, 4(3&4). http://www.agbioforum.org/.

Palumbi, Stephen R. 2001. Humans as the World's Greatest Evolutionary Force. *Science* 293(5536):1786–90, 7 September.

Pang, Jenny W. Y., James D. Heffelfinger, Greg J. Huang, Thomas J. Benedetti, and Noel S. Weiss. 2002. Outcomes of Planned Home Births in Washington State: 1989–1996. *Obstetrics & Gynecology* 100(2):253–59. August.

Park, Robert L. 2002. Alternative Medicine and the Laws of Physics. *CSIOCP* (Committee for the Scientific Investigation of Claims of Paranormal). Csicop.org/si/9709/park.html.

Pawson, Tony. 2002. Foreward. In *Genes and Signals* by Mark Ptashne and Alexander Gann, pp. xv–xvi. Cold Springs Harbor, NY: Cold Springs Harbor Laboratory Press.

Pence, Gregory E. 2002. *Designer Food: Mutant Harvest or Breadbasket to the World?* Lanham, MD: Rowman & Littlefield.

Pert, Candace B. 1997. *Molecules of Emotion: Why You Feel The Way You Feel* (foreword by Deepak Chopra). New York: Scribner.

Peukert, Detlev. 1988. The Genesis of the "Final Solution" from the Spirit of Science, Conference on Re-Evaluating the "Third Reich": Interpretations and Debates, University of Pennsylvania, April 8–10, 1988.

_____. 1992. *The Weimar Republic: The Crisis of Classical Modernity* (translated by Richard Deveson). New York: Hill and Wang.

_____. 1993. The Genesis of the "Final Solution" from the Spirit of Science. In *Reevaluating the Third Reich,* edited by Thomas Childers and Jane Caplan, pp. 234–52. New York: Holmes & Meier.

Peukert, Detlev J. K. 1987. *Inside Nazi Germany: Conformity, Opposition, and Racism in Everyday Life* (translated by Richard Deveson). New Haven: Yale University Press.

PHS (Public Health Services). 1976. *Child in America.* Rockville, MD: U.S. Department of Health, Education and Welfare, Public Health Services.

Pigott, Nick, Vas Novelli, Suneel Pooboni, Richard Firmin, and Allan Goldman. 2002. The Importance of Herd Immunity Against Infection. *The Lancet* 360(9333), 24 August.

Pleasants, John M., Richard L. Hellmich, Galen P. Dively, Mark K. Sears, Diane E. Stanley-Horn, Heather R. Mattila, John E. Foster, Thomas L. Clark, and Gretchen D. Jones. 2001. Corn Pollen Deposition on Milkweeds in And Near Cornfields. *PNAS* (Proceedings of the National Academy of Sciences USA) 98, 11 September.

Pois, Robert A. 1986. *National Socialism and the Religion of Nature.* New York: St. Martin's Press.

Polkinghorne, John. 2002. Thoughts of a Rationalist: A Review of *Facing Up: Science and Its Cultural Adversaries* by Steven Weinberg, Cambridge, MA: Harvard University Press. *Nature* 416(6881):583, 11 April.

Pollack, Andrew. 2001. Scientists Are Adding Letters to Life's Alphabet. *New York Times*, 24 July.

_____. 2002. Drug That Blocks Blood Flow Slows Tumor Growth in Trial. *New York Times*, 20 May.

Pollan, Michael. 1994. Against Nativism: Horticultural Formalism May Be Out, But the New American Garden Free of Foreign Flora and Human Artifice, A as Natural As Its Advocates Claim. *New York Times Magazine*, 52–55, 15 May.

_____. 2002. An Animal's Place. *New York Times Magazine*, 10 November.

Portugal, Franklin H., and Jack S. Cohen. 1977. *A Century of DNA: A History of the Discovery of the Structure and Function of the Genetic Substance.* Cambridge, MA: MIT Press.

Postrel, Virginia. 2001. Criminalizing Science: Leading Thinkers and Commentators Respond to a Left-Right Alliance to Outlaw "Therapeutic Cloning" and Stigmatize Genetic Research. *Reason*, November.

Potter, Angela. 2002. Maryland Picks Poison to Wipe Out Snakeheads. *Houston Chronicle* (Associated Press), 18 August.

Powell, Kendall. 2002. Rice Genes Removed: New Method Aids Gene Study in Cereal Plants. *Nature* online, 9 September.

Proctor, Robert. 1988. *Racial Hygiene: Medicine Under the Nazis.* Cambridge: Harvard University Press.

_____. 1999. *The Nazi War on Cancer.* Princeton: Princeton University Press.

_____. 2002. The Nazi Campaign Against Tobacco. In *Medicine and Medical Ethics in Nazi Germany: Origins, Practices, Legacies,* edited by Francis R. Nicosia and Jonathan Huener, pp. 40–58. New York: Berghahn Books.

Quist, David, and Ignacio H. Chapela. 2001. Transgenic DNA Introgressed into Traditional Maize Landraces in Oaxaca, Mexico. *Nature* 414(6863):541–43, 29 November.

_____. 2002. Biodiversity Communications Reply: Suspect Evidence of Transgenic Contamination/Maize Transgene Results in Mexico Are Artifacts (followed by an editorial note). Published online: 4 April 2002. *Nature* 416(6881):602, 11 April.

Rabkin, Yakov M. 1987. Technological Innovation in Science: The Adoption of Infrared Spectroscopy by Chemists. *Isis* 78(291):31–54, March.

Raichle, Marcus E., and Debra A. Gusnard. Appraising the Brain's Energy Budget. *PNAS* (Proceedings of the National Academy of Sciences) 99(16):10237–39, 6 August.

Rajkowski, Kathleen T., Glen Boyd, and Donald W. Thayer. 2003. Irradiation D-Values for *Escherichia coli* O157:H7 and *Salmonella* spp. on Inoculated Broccoli Seeds and Effects of Irradiation on Broccoli Sprout Keeping Quality and Seed Viability. *Journal of Food Protection* 66(5):760,766, May.

Randerson, James. 2003. Early Chefs Left Indelible Mark On Human Evolution. *New Scientist* 177(2387):37, 22 March.

Randhawa, Mohindar Singh. 1983. *A History of Agriculture in India.* New Delhi: Indian Council of Agricultural Research.Rangan, Haripriya. 1993. Romancing the Environment: Popular Environmental Action in Garhwal Himalayas. In *In Defense of Livelihood: Comparative Studies on Environmental Action,* edited by John Friedmann and Haripriya Rangan, pp. 155–81. West Hartford, Conn.: Kumarian Press.

Rangan, Haripriya. 2000. *Of Myths and Movements: Rewriting Chipko into Himalayan History.* London, New York: Verso.

Rasmussen, Nicolas, 1997. *Picture Control: The Electron Microscope and the Transformation of Biology in America, 1940–1960.* Stanford, Calif.: Stanford University Press.

Rauschning, Herman. 1940. *Gesprache Mit Hitler.* New York: Europa Verlag.

Raven, Peter. H. 2003. The Environmental Challenge. Presented at the Natural History Museum, London, 22 May.

Reichel, Ronald R. 1983. *Of a Fire on the Earth: The Struggle Against the Dehumanization of Technology.* Albuquerque: University of New Mexico Thesis (Ph.D.).

Rhodes, James M. 1980. *The Hitler Movement: A Modern Millenarium Revolution*. Stanford: Hoover Institution Press.

Richards, Robert J. 2002. *The Romantic Conception of Life Science and Philosophy in the Age of Goethe*. Chicago, IL: University of Chicago Press.

Rifkin, Jeremy. 2001. This Is The Age Of Biology: Left And Right Are Finding Common Ground In Opposition To A Utilitarian View Of Life. *The Guardian* (London), 28 July.

Riley, James C. 2001. *Rising Life Expectancy: A Global History*. Cambridge, New York: Cambridge University Press.

Roane, Kit. 2002. Ripe for Abuse: Farmworkers Say Organic Growers Don't Always Treat Them as Well as They Do Your Food. *US News & World Report* 22 April.

Rocheford, Torbert R., Jeffrey C. Wong, Cem O. Egesel, and Robert J. Lambert. 2002. Enhancement of Vitamin E Levels in Corn. *Journal of the American College of Nutrition* (Special Supplement—*The Future of Food and Nutrition With Biotechnology*) 21(3):191S–198S, June.

Rohter, Larry. 2003. Antiglobalization Forum to Return to a Changed Brazil. *New York Times*, 20 January.

Ronald, Pamela, and Hei Leung. 2002. The Rice Genome: The Most Precious Things Are Not Jade and Pearls. . . . *Science* 296(5565):58–59, 5 April.

Rosenberg, Noah A., Jonathan K. Pritchard, James L. Weber, Howard M. Cann, Kenneth K. Kidd, Lev A. Zhivotovsky, and Marcus W. Feldman. 2002. Genetic Structure of Human Populations. *Science* 298(5602):2381–85, 20 December.

Rusbridger, Alan. 2002. Drama Taps into GM Debate. *BBC World Service* on-line, 7 June.

Russell, Edmund P., III. 2001. *War and Nature: Fighting Humans and Insects with Chemicals from World War I to Silent Spring*. Cambridge University Press.

Sachs, Jeffrey D. 2001. *Macroeconomics and Health: Investing in Health for Economic Development* (Report of the Commission on Macroeconomics and Health, chaired by Jeffrey D. Sachs). Geneva: World Health Organization.

Sala-i-Martin, Xavier. 2002. *The Disturbing "Rise" of Global Income Inequality*. Washington, D.C.: NBER (National Bureau for Economic Research) Working Paper No. w8094, April.

Salleh, Anna. 2002. Mexican Maize Madness. *ABC* (Australian Broadcasting Corporation) *Science Program, Lab Online*, July. abc.net.au/science/slab/mexicanmaize/default.htm.13.

Sanchez, Ivelisse, Christian Mahlke, and Junying Yuan. 2003. Pivotal Role of Oligomerization in Expanded Polyglutamine Neurodegenerative Disorders. *Nature* 421(6921):373–79, 23 January.

Sanders, John H. 2001. Agriculture: Another World Food Scare? A Review of *Food's Frontier: The Next Green Revolution* by Richard Manning [North Point (Farrar, Straus, and Giroux), New York, 2000]. *Science* 291(5509):1707–1708, 2 March.

Santerre, Charles R., and Krisanna L. Machtmes. The Impact of Consumer Food Biotechnology Training on Knowledge and Attitude. *Journal of the American College of Nutrition* (Special Supplement—*The Future of Food and Nutrition With Biotechnology*) 21(3):174S–7S, June.

Sassower, Raphael. 1997. *Technoscientific Angst: Ethics + Responsibility.* Minneapolis: University of Minnesota Press.

Sax, Boria. 2000. *Animals in the Third Reich: Pets, Scapegoats, and the Holocaust* (foreword by Klaus P. Fischer). New York: Continuum.

Schama, Simon 1995. *Landscape and Memory.* New York: A.A. Knopf, Distributed by Random House.

Schiff, Michel. 1995. *The Memory of Water: Homoeopathy and the Battle of Ideas in the New Science.* London: Thorsons.

Scruton, Roger. 2000. Herbicide, Pesticide, Suicide: Seed Merchants Prosper and Farmers Wither; That's the Truth of Global Agribusiness. *Financial Times* (London), 6 June.

Sears, Mark K., Richard L. Hellmich, Diane E. Stanley-Horn, Karen S. Oberhauser, John M. Pleasants, Heather R. Mattila, Blair D. Siegfried, and Galen P. Dively. 2001. Impact of BT Corn Pollen on Monarch Butterfly Populations: A Risk Assessment. *PNAS* (Proceedings of the National Academy of Sciences USA) 98, 11 September.

Seidelman, William E. 1996. Nuremberg Lamentation: for the Forgotten Victims of Medical Science. *BMJ* (*British Medical Journal*) 313(7070):1463–67, 7 December.

Seife, Charles. 2000. Boomerang Returns with Surprising News. *Science* 288(5466):595, 28 April.

Sengupta, Somini. 2003. When Do-Gooders Don't Know What They're Doing. *New York Times*, 11 May.

Sephton, Mark A. 2001. Meteoritics: Life's Sweet Beginnings? *Nature* 414(6866):857–58, 20/27 December.

Sephton, Mark A., and I. Gilmour. 2001. Compound-specific Isotope Analysis of the Organic Constituents in Carbonaceous Chondrites. *Mass Spectrometry Reviews* 20(3):111–20, May.

Serageldin, Ismail. 2002. The Rice Genome: World Poverty and Hunger—the Challenge for Science. *Science* 296(5565):54–58, 5 April.

Sevenier, Robert, Ingrid M. van der Meer, Raoul Bino, and Andries J. Koops. 2002. Increased Production of Nutriments by Genetically Engineered Crops. *Journal of the American College of Nutrition* (Special Supplement—*The Future of Food and Nutrition With Biotechnology*) 21(3):199S–204S, June.

Shelton, Anthony M., and Mark K. Sears. 2001. The Monarch Butterfly Controversy: Scientific Interpretations of a Phenomenon. *The Plant Journal* 27(6):483–88, October.

Shelton, Anthony M., and Richard T. Roush. 1999. False Reports and the Ears of Men. *Nature Biotechnology* 17(9):832, September.

Shiva, Vandana. 1991. *The Violence of the Green Revolution: Third World Agriculture, Ecology, and Politics.* London: Zed Books.

————. 2000. BBC Reith Lectures 2000. *BBC* online, 12 May.

Shock, Everett. 2002. Astrobiology: Seeds of Life? *Nature* 416(6879): 380–81, 28 March.

Shouse, Ben. 2002. Genetically Modified Food: TV Drama Sparks Scientific Backlash. *Science* 296(5575):1948–49, 14 June.

Sicherer, Scott H. 2002. Food Allergy. *The Lancet* 360(9334):701–10, 31 August.

Siegel, Jay S. 2002. Chemistry: Shattered Mirrors. *Nature* 419(6905):346–47, 26 September.

Slakey, Francis. 1993. When the Lights of Reason Go Out. *New Scientist* 139(1890):49–50, 11 September.

Smetacek, Victor. 2002. Balance: Mind-grasping Gravity. *Nature,* 415(6871):481, 31 January.

Smil, Vaclav. 2000. *Feeding the World: A Challenge for the Twenty-first Century.* Cambridge: MIT Press.

————. 2001. *Enriching the Earth: Fritz Haber, Carl Bosch, and the Transformation of World Food.* Cambridge: MIT Press.

————. 2002. *The Earth's Biosphere: Evolution, Dynamics, and Change.* Cambridge, MA: MIT Press.

Smith, Barry. 2002. Nitrogenase Reveals Its Inner Secrets. *Science* 297 (5587): 1654–55, 6 September.

Smuts, Jan Christian. 1926. *Holism and Evolution.* London: Macmillan & Co.

Solomon, Ethan B., Catherine J. Potenski, and Karl R. Matthews. 2002. Effect of Irrigation Method on Transmission to and Persistence of *Escherichia coli* 0157: H7 on Lettuce. *Journal of Food Protection* 65(4):673–76, April.

Solomon, Ethan B., Sima Yaron, and Karl R. Matthews. 2002. Transmission of *Escherichia coli* O157:H7 from Contaminated Manure and Irrigation Water to Lettuce Plant Tissue and Its Subsequent Internalization. *Applied and Environmental Microbiology* 68(1):397–400, January.

Soyinka, Wole. 1995. Why the General Killed, 20 November. http://www.prairienet.org/acas/soyinka.html.

Spielvogel, Jackson, and David Redles. 1986. Hitler's Racial Ideology: Content and Occult Sources. In Henry Friedlander and Sybil Milton, eds., *Simon Wiesenthal Center Annual* (Los Angeles) 3:229–41.

Srinivasan, Gayathri, Carey M. James, and Joseph A. Krzycki. 2002. Pyrrolysine Encoded by UAG in Archaea: Charging of a UAG- Decoding Specialized tRNA *Science* 296(5572):1459–62, 24 May.

Stanley-Horn, Diane E., Galen P. Dively, Richard L. Hellmich, Heather R. Mattila, Mark K. Sears, Robyn Rose, Laura C. H. Jesse, John E. Losey, John J. Obrycki, and Les Lewis. 2001. Assessing the Impact of Cry1AB-expressing Corn Pollen on Monarch Butterfly Larvae in Field Studies. 2001. *PNAS* (Proceedings of the National Academy of Sciences USA) 98, 11 September.

Steiner, Rudolf. 1958. *Agriculture: A Course of Eight Lectures* (translated by George Adams). London: Biodynamics Association.

Sterner, Robert W., and James J. Elser. 2002. *Ecological Stoichiometry: The Biology of Elements from Molecules to the Biosphere.* Princeton, NJ: Princeton University Press.

Stokstad, Erik. 2001. Entomology: First Light on Genetic Roots of Bt Resistance. *Science* 293(5531):778, 3 August.

————. 2002. Organic Farms Reap Many Benefits. *Science* 296(5573): 1589, 31 May.

STPP (Society of Toxicology Position Paper). 2002. The Safety of Genetically Modified Foods Produced Through Biotechnology. Society of Toxicology Position Paper, 25 September.

Surridge, Christopher. 2002. Agricultural Biotech: The Rice Squad. *Nature* 416(6881):576–78, 11 April.

Susman, Robert W. 1987. Morpho-physiological Analysis of Diets: Species-specific Dietary Patterns in Primates and Human Dietary Adaptations. In *The Evolution of Human Behavior: Primate Models*, edited by Warren G. Kinzey, pp. 151–82. New York: State University of New York Press.

Tanford, Charles, and Jacqueline Reynolds. 2001. *Nature's Robots: A History of Proteins.* New York: Oxford University Press.

Tanne, Janice Hopkins. 2002. The Epidemic That Never Was. *BMJ (British Medical Journal)* 325(7367):782, 5 October.

Terada, Rie, Hiroko Urawa, Yoshishige Inagaki, Kazuo Tsugane, and Shigeru Iida. 2002. Efficient Gene Targeting by Homologous Recombination in Rice. *Nature* online, 9 September.

Teresi, Dick. 2000. It's Been Hell on Earth: A Historian Chronicles Our Assault on the Planet in the Last Hundred Years, a review of *Something New Under the Sun: An Environmental History of the Twentieth-Century World* by J.R. McNeill. New York: W.W. Norton. *New York Times*, 25 June.

Toby, Sidney. 2000. Acid Test Finally Wiped Out Vitalism, And Yet. . . . *Nature* 408(6814):767, 14 December.

Toy, Vivian S. 2002. Long Island Study Sees No Cancer Tie to Pesticides. *New York Times*, 6 August.

Travis, Anthony. 1989. Science as a Receptor of Technology: Paul Erhlich and the Synthetic Dye Industry. *Science in Context* 3(2):383–408, Autumn.

TTOI (The Times of India). 2002. Indoor Air Pollution Taking Its Toll in India. *The Times of India*, 9 February.

Tyer, Brad. 2000. RiceTec Paddy Whack. *Houston Press*, 23 November. http://www.houstonpress.com/issues/20001123/feature2.html/1/index.html.

UNDP (United Nations Development Programme). 2001. *Human Development Report 2001: Making New Technologies Work For Human Development.* New York: Oxford University Press, Published for the United Nations Development Programme.

UN Wire. 2002. Indoor Stoves: WHO Reports Indoor Air Pollution Is Killing Millions. UN Wire Issue for 3 September.

USDA (United States Department of Agriculture). 2002. The National List of Allowed and Prohibited Substances, Washington, D.C.: National Organic Program, Agricultural Marketing Service, United States Department of Agriculture. http://www.ams.usda.gov/nop/NationalList/FinalRule.html.

USFDA (United States Food and Drug Administration). 2002. Modified Soybeans May Be Less Allergenic. Washington, DC: United States Food and Drug Administration, Agricultural Research Service, ARS News Service, 3 September.

Vandenbroucke, J. P., and A. J. M. de Craen. 2001. Alternative Medicine: A "Mirror Image" for Scientific Reasoning in Conventional Medicine. *Annals of Internal Medicine* 135(7):597–13, 2 October.

Veblen, Thorstein B. [1914] 1964. *The Instinct of Workmanship and the State of the Industrial Arts.* New York: Agustus Kelley.

Veggeberg, Scott. 2002. Fighting Cancer with Angiogenesis Inhibitors: Clinical Trials Suggest Silver Bullet May Be Losing Its Luster. *The Scientist* 16(11):41, 27 May.

Wade, Nicholas. 2002. Gene Study Identifies 5 Main Human Population. *Science* 298(5602):2381–85, 20 December.

_____. 2003. Drug Shows Promise in at Least 4 Cancers. *New York Times*, 27 March.

Waggoner, P. E., and J. H. Ausubel. 2002. A Framework for Sustainability Science: A Renovated IPAT Identity. *PNAS* (*Proceedings of the National Academy of Sciences*) 99(12):7860–65, 11 June.

Walker, Alan. 1984. Extinctions in Hominid Evolution. In *Extinctions,* edited by Matthew E. Nitecki, pp. 119–52. Chicago: University of Chicago Press.

Walston, Oliver. Organic Food Is No Safer Than The Ordinary Cheaper Kind, 2002. *Daily Telegraph* (London) 28 March.

Wang, Lei, Ansgar Brock, Brad Herberich, and Peter G. Schultz. 2001. Expanding the Genetic Code of *Escherichia coli. Science* 292:5516 (498–500), 20 April.

Watson, J. D., and F. H. C. Crick. 1953a. Molecular Structure of Nucleic Acids. *Nature* 171(4356):737–38, 25 April.

————. 1953b. Genetical Implications of the Structure of Deoxyribonucleic Acid. *Nature* 171(4361):964–67, 30 May.

Weatherall M. W. 1996. Making Medicine Scientific: Empiricism, Rationality, and Quackery in Mid-Victorian Britain. *Social History of Medicine* 9(2):175–94, August.

Webb, James. 1976. *The Occult Establishment.* La Salle, IL: Open Court Publishing Co.

————. 1980. *The Harmonious Circle: The Lives and Work of G. I. Gurdjieff, P. D. Ouspensky, and Their Followers.* London: Thames and Hudson.

Weiner, Tim. 2002. Growing Poverty Is Shrinking Mexico's Rain Forrest. *New York Times,* 8 December.

Weisskopf, M. C., L. P. Van Speybroeck, G. P. Garmire, S. S. Murray, A. Brinkman, T. Gunsing, J. Kaastra, R. van der Meer, R. Mewe, F. Paerels, J. van Rooijen, H. Brauninger, V. Burwitz, G. Hartner, G. Kettenring, P. Predehl, D. Dewey, H. Marshall, J. Chappell, J. Drake, O. Johnson, A. Kenter, R. Kraft, G. Meehan, P. Ratzlaff, B. Wargelin, M. Zombeck, and C. Canizares. 2000. Session 96: First Results from the Chandra X-ray Observatory. *195th Meeting of the American Astronomical Society*, Atlanta, GA, 11–15 January.

Wells, Matt. 2002. BBC Defends Upcoming Drama About Potential Dangers of GM Crops. *The Guardian* (London), 31 May.

White, Caroline. 2002. Life Expectancy is Consistently Underestimated, Say Researchers. *BMJ (British Medical Journal)* 324(7348):1173, 18 May.

Whorton, James C. 2002. *Nature Cures: The History of Alternative Medicine in America.* Oxford, New York: Oxford University Press.

Wiener, Norbert. 1989. *The Human Use of Human Beings: Cybernetics and Society.* London: Free Association.

Wilford, John Noble. 1998. Peering Back in Time, Astronomers Glimpse Galaxies Aborning: Long, Long Ago and Far, Far Away. *New York Times*, 20 October.

————. 2000. Craft Is Demystifying Milky Way's Sizzling Halo of Gas. New York Times, 13 January.

Wills, Christopher, and Jeffrey L. Bada. 2000. *The Spark of Life: Darwin and the Primeval Soup.* Cambridge, MA.: Perseus.

Windschuttle, Keith. 1997. *The Killing of History: How Literary Critics and Social Theorists Are Murdering Our Past.* New York: The Free Press.

Wistrich, Robert S. 2001. *Hitler and the Holocaust.* New York: Modern Library.

Wolfe, Robert M., and Lisa K. Sharp. 2002. Anti-vaccinationists Past and Present. *BMJ (British Medical Journal)* 325(7361):430–32, 24 August.

Wolin, Richard. 1990. *The Politics of Being: The Political Thought of Martin Heidegger.* New York: Columbia University Press.

_____. 2001. *Heidegger's Children: Hannah Arendt, Karl Lowith, Hans Jonas, and Herbert Marcuse.* Princeton, NJ: Princeton University Press.

Wolin, Richard, editor. 1993. *The Heidegger Controversy: A Critical Reader.* Cambridge, MA: MIT Press.

Wolschke-Bulmahn, Joachim. 1992. The Fear of the New Landscape: Aspects of the Perception of Landscape in the German Youth Movement Between 1900 and 1933 and Its Influence on Landscape Planning. *Journal of Architectural and Planning Research* 9(1):33–47, Spring.

_____. 1994. Ethics and Morality. Questions in the History of Garden and Landscape Design: A Preliminary Essay. *Journal of Garden History* 14(3):140.

_____. 1997a. Garden-Variety Xenophobia Source. *Harper's* 294 (1761):15.

_____. 1997b. The Nationalism of Nature and the Naturalization of the German Nation: "Teutonic" Trends in Early Twentieth-Century Landscape Design. In *Nature and Ideology: Natural Garden Design in the Twentieth Century,* edited by Joachim Wolschke-Bulmahn, pp. 187–219. Washington, DC: Dumbarton Oaks Research Library and Collection.

Wrangham, Richard W. 2001. Out of the Pan, Into the Fire: How Our Ancestor's Evolution Depended on What They Ate. In *Tree of Origin: What Primate Behavior Can Tell Us About Human Social Evolution,* edited by Frans B. M. de Waal, pp. 119–43. Cambridge, MA: Harvard University Press.

Wrangham, Richard W., and Nanvy Lou Conklin-Britain. 2003. Cooking as a Biological Trait. *Comparative Biochemistry and Physiology: Part A— Molecular and Integrative Physiology* 134 (in press).

Wrangham, Richard W., James Holland Jones, Greg Laden, David Pilbeam, and Nanvy Lou Conklin-Britain. 1999. The Raw and the Stolen: Cooking and the Ecology of Human Origins. *Current Anthropology* 40(1):567–77, 586–94, December.

Yu, Ju, Songnian Hu, Jun Wang, Gane Ka-Shu Wong, Songgang Li, Bin Liu, Yajun Deng, Li Dai, Yan Zhou, Xiuqing Zhang, Mengliang Cao, Jing Liu, Jiandong Sun, Jiabin Tang, Yanjiong Chen, Xiaobing Huang, Wei Lin, Chen Ye, Wei Tong, Lijuan Cong, Jianing Geng, Yujun Han, Lin Li, Wei Li, Guangqiang Hu, Xiangang Huang, Wenjie Li, Jian Li, Zhanwei Liu, Long Li, Jianping Liu, Qiuhui Qi, Jinsong Liu, Li Li, Tao Li, Xuegang Wang, Hong Lu, Tingting Wu, Miao Zhu, Peixiang Ni, Hua Han, Wei Dong, Xiaoyu Ren, Xiaoli Feng, Peng Cui, Xianran Li, Hao Wang, Xin Xu, Wenxue Zhai, Zhao Xu, Jinsong Zhang, Sijie He, Jianguo Zhang, Jichen Xu, Kunlin Zhang, Xianwu Zheng, Jianhai Dong,

Wanyong Zeng, Lin Tao, Jia Ye, Jun Tan, Xide Ren, Xuewei Chen, Jun He, Daofeng Liu, Wei Tian, Chaoguang Tian, Hongai Xia, Qiyu Bao, Gang Li, Hui Gao, Ting Cao, Juan Wang, Wenming Zhao, Ping Li, Wei Chen, Xudong Wang, Yong Zhang, Jianfei Hu, Jing Wang, Song Liu, Jian Yang, Guangyu Zhang, Yuqing Xiong, Zhijie Li, Long Mao, Chengshu Zhou, Zhen Zhu, Runsheng Chen, Bailin Hao, Weimou Zheng, Shouyi Chen, Wei Guo, Guojie Li, Siqi Liu, Ming Tao, Jian Wang, Lihuang Zhu, Longping Yuan, and Huanming Yang. 2002. A Draft Sequence of the Rice Genome (Oryza sativa L. ssp. indica). *Science* 296(5565):79–92, 5 April.

Zangerl, A. R., D. McKenna, C. L. Wraight, M. Carroll, P. Ficarello, R. Warner, and M. R. Berenbaum. 2001. Effects of Exposure to Event 176 *Bacillus thuringiensis* Corn Pollen on Monarch And Black Swallowtail Caterpillars Under Field Conditions. *PNAS* (Proceedings of the National Academy of Sciences USA) 98, 11 September.

Zimmerman, Michael E. 1990. *Heidegger's Confrontation with Modernity: Technology, Politics, and Art.* Bloomington, IN: Indiana University Press.

Zimmermann, Erich W. 1951. *World Resources and Industries: A Functional Appraisal of the Availability of Agricultural and Industrial Materials.* New York: Harper & Brothers.

Index

absolutist assertions, mainsprings of ideology, 84
adnine (A), DNA chemical unit, 99–100
aflatoxins, threat to human life, 144
agricultural biotechnologists, thioredoxin protein production, 109
agriculture, importance to human life, 119–120
alanine, first amino acid, 17
allergenic proteins
    amino acids, 108–114
    gene silencing, 109
amateur astronomers, accomplishments, 164–165
amines, food sensitivity cause, 113
amino acids
    20 building blocks of proteins, 98–99
    alanine, 17
    allergenic proteins, 108–114
    defined, 98
    DNA chemical units (bases), 99–100
    nitrogen, 98–100
    nonstandard, 99
    origins of life, 6–8
ammonia, industrial synthesis, 49
angiogenesis inhibitors, "smart" pharmaceuticals, 23
aniline dyes
    mauve decades, 14–15
    sulfa development, 16
animal rights, Nazi support, 72–73
anthropocentrism, Nazi opposition to, 71
antibiotics, development history, 19
antiscience, adverse consequences, 85
anti-tryspin factor (ATF), soybeans, 142
astronomers, amateur versus professional, 164, 165
astronomy, 163–167
astronomy, Dobsonian telescope, 164–165
ATF (anti-tryspin factor), soybeans, 142
atmospheric nitrogen, industrial synthesis of ammonia, 49

autism, MMR vaccine concerns, 50–51

Bacillus thuringinesis (Bt) corn, enzyme activated protein, 109–110
barley, T gene, 115
big bang, background radiation, 164
biodynamic organic farming, nonuse of man-made fertilizer, 50
bioengineered Bt (Bacillus thuringinesis) corn, 109–110
biology, reductionism form, 21
biotechnology, origins of, 33
BOOMERANG, microwave telescope, 167
breast cancer
    Herceptin treatment, 23
    probable causes, 161–162
breast feeding, breast cancer correlation, 161–162
Bt (Bacillus thuringinesis) corn
    corn borer as target, 120
    enzyme activated protein, 109–110
    monarch butterfly, 117–118

C.C.D. (light sensing charge-coupled device), 165
C4 photosynthesis, two-stage process, 130
cancer, molecular targeting, 23
cancers, statistical data, 145–146
cells
    chromosomes, 15–16
    discovery of, 12–13
chemistry, reductionism form, 21, 23–24
Chipko (tree huggers) movement, model for green ideologies, 136–137
chirality, amino acid property, 7
Chol Indians, 158
chromosomes, cell material, 15–16
CIGAR (Consultative Group on International Agricultural Research), 121
CIMMYT (International Center for the Improvement of Maize and Wheat), 121

Claviceps purpurea, grain fungus, 18
CMOS (complementary metal-oxide
semiconductor) chips, astronomy, 165
commercial plants, inbred resistance genes,
115
compulsory immunization, exemptions
based on conscience, 51
Congress of Racial Equality (CORE), 149
consciousness, reductionism forms, 21
Consultative Group on International
Agricultural Research (CIGAR), 121
containers, food processing technology, 142
CORE (Congress of Racial Equality), 149
corn borer, target of Bt corn, 120
cosmic ice theory, Nazi lunacy, 76–78
cytosine (C), DNA chemical unit, 99–100

Darwinian revolution, overcoming vitalist
beliefs, 11–12
DDT, equated to chemical warfare, 60–61
deconstructionism, origins in Nazism, 55
democratic society, importance of
compromise, x
deoxyribose nucleic acid. See DNA
diamondback moths, Bt resistance,
115–116
diseases, germ theory, 15
D-isomers, sugars, 7
DNA (deoxyribose nucleic acid)
chemical units (bases), 99–100
helix structure revealed, 29–32
Darwinian classification conformation,
11–12
DNA research
helix structure revealed, 29–32
pairing relationships, 29
reductionism form, 25
shared genome percentages, 25
what a "gene" is, 27–32
Dobsonian telescope, 164, 165
domesticated plants, weakened defense
mechanisms, 114–115

E. coli, infestation methods, 145
earth, heterotrophic life forms, 100–102
electromagnetic spectrum, radio waves, 163
ELISA (Enzyme-Linked Immunosorbent
Assay), 121
entropy, second law of thermodynamics,
4–5
Enzyme-Linked Immunosorbent Assay
(ELISA), 121
enzymes, name origins, 13
equilibrium, corrective feedback, 43
ergometrine, development history, 18
ergot
grain infection, 18
life-saving uses, 18–19

eugenics, spectre of Nazism raised, 55
evolution, development concepts, 44–45

farmers, transgenic crop benefits, 129–130
ferments, vitalist beliefs, 13
fertilizers, synthetic nitrogen, 103–108
fire
charcoal uses, 142
food processing role, 138–140
mandatory in northern climate food
processing, 141
first law of thermodynamics, conservation
of matter and energy, 4
food processing
container impact, 142
essential component of food safety,
138–144
technologies, 142
vitamin C reduction, 143
food supply
contamination concerns, 144
E. coli infestation methods, 145
importance in species survival, 140
limits-to-growth theory, 96
microbes, 145
foods, organic versus conventional,
109–114
Frankenfears, Internet scare scenarios,
85–87
fungi
Claviceps purpurea, 18
penicillin, 17–18

gemules, Darwin's mechanism for
inheritance, 28
genes
Hope, 115
inbred in commercial plants, 115
LR-34, 115
mechanisms for inheritance, 27–32
structural versus regulator, 31–32
genetics, essential to twentieth-century
agriculture advances, xvii
Gleevec, chronic myeloid leukemia
treatment, 23
globalization, opponents, 153–157
glutamates, food sensitivity cause, 113
GM crops
Internet Frankenfears, 85–87
precautionary principle, 83–85
grains, Claviceps purpurea, 18
gram stain, development of, 16
green movement, Nazism similarities,
53–55
green revolution, science and technology
success, xvi–xvii
guanine (G), DNA chemical unit,
99–100

Hale-Bopp comet, discovery, 165
Hamilton, David, x–xii
Herceptin, breast cancer treatment, 23
heterotrophic life forms, early earth life, 100–102
holism
    development concepts, 44–45
    purity of the past belief, 56–58
Holocaust
    compared to the practice of medicine, 55–56
    logical outcome of the logic of modern science belief, 67–68
homeopathy
    astral body belief, 50
    dilution notations, 46–47
    ISIS (Institute of Science in Society), 48
    Law of Infinitesimals, 46
    Law of Similia, 46
    principles of, 46
    progressively less harm approach, 47–48
    pseudoscience currents, 41–42
    pseudoscientific term use, 50
    randomized test results, 48
    re-enchantment of science, 45–48
    substance "memory" retention, 47
    versus reductionist medicine, 58–59
    "what doesn't kill me makes me stronger", 51
Hope gene, stem rust prevention, 115
Hubble telescope, discoveries, 166
human evolution, development of balance mechanisms, 43
human machine, scientific nutrition's origins, 3–5

ideas, responsibility of those who hold them, 78–80
immunizations
    herd immunity protection level, 51–52
    Waldorf School opposition, 51
incremental change, virtue of, x–xi
Industrial Revolution, effect on population growth, 96
inquiry, unity/beauty of, 5–6
insects, Bt resistant strains, 115–118
Institute of Science in Society (ISIS), homeopathy supporter, 48
International Center for the Improvement of Maize and Wheat (CIMMYT), 121
Internet, GM crop Frankenfears, 85–87
inverse Polymerase Chain Reaction (iPCR), false positives, 120–121
iPCR (inverse Polymerase Chain Reaction), false positives, 120–121
Iressa, lung cancer treatment, 23

IRRI (International Rice Research Institute), 121
ISIS (Institute of Science in Society), homeopathy supporter, 48

Klebsiella planticola, Internet Frankenfears, 86–87

landscape, Nazi ideology, 70–71
Law of Infinitesimals, homeopathy principle, 46
Law of Similia, homeopathy principle, 46
leaf rust, LR-34 gene as limiter, 115
Liebig's Law of Minimum, 108
L-isomers, amino acids, 7
logocentrism, origins in Nazism, 55
Long Island Breast Cancer study, 161–162
LR-34 gene, leaf rust limiter, 115

maize, C4 photosynthesis, 130
mathematics, zero and positional notation, 37
meats, essential for the emergence of humans, 140–141
meteorites, polyhydroxylated compounds, 6–8
Mexican maize landraces, evidence of transgenes, 120–124
microbes, presence in food supply, 145
microorganisms, last bastion of vitalist position, 41–42
microscopes
    cell discovery tool, 12–13
    gram stain, 16
    phase contrast, 16
microwave telescope, BOOMERANG, 167
minerals, soil fertilization recognition, 11
MMR vaccine, autism concerns, 50–51
molecular biology
    biotechnology development role, 33
    essential to twentieth-century agriculture advances, xvii
    foundations, 14
    overcoming vitalist beliefs, 12
    reductionism form, 25
    RNA/DNA sequencing, 32–33
molecular targeting, peptide "zip codes", 23
monarch butterfly, Bt corn, 117–118
morality, vitalism relationships, 42–43
multiculturalism
    extreme in parochialism, 38
    "local ways of knowing" superiority myth, 34–38

natural medicine, Nazi ideology, 59–61
nature, agriculture relationship, 119–120
Nazism
    animal rights, 72–73

Nazism *(continued)*
    concept of humanity as biological
        nonsense, 71
    cosmic ice theory, 76–78
    cost of rejected knowledge, 53–63
    environmental deconstruction, 54–55
    green movement similarities, 53–55
    Holocaust as logical outcome of modern
        science belief, 67–68
    Holocaust compared to the practice of
        medicine, 55–56
    homeopathy versus reductionist
        medicine, 58–59
    landscape beliefs, 70–72
    modern science in disfavor, 73–75
    natural medicine, 59–61
    occult beliefs, 73–78
    opposition to humans dominating
        nature, 66–67
    organic agriculture, 59–61
    postmodernist beliefs, 68–69
    purity of the past belief, 56–58
    racial hygiene, 72–73
    vegetarianism, 59–61
    war of extermination against alien
        plants, 71
Newton's laws, unification of terrestrial and
    celestial mechanics, 27
NGOs (non-governmental organizations)
    annual budgets, 126
    feeding the hungry as agenda element,
        127
    government funding, 126–127
    social protest, 153–158
    unfolding threat to biodiversity, 105
nitrogen
    amino acids, 98–100
    early earth life forms, 100–102
    industrial synthesis of ammonia, 49
    life supporting resource, xviii
    organic agriculture, 97–98
    reductionist science, 103
    synthetic fertilizer, 103–108
nucleic acids
    name origins, 14
    transforming factor of inheritance,
        28–29

occult belief systems, Nazism, 73–78
Ockham's razor, origin of reductionism, 4
Ogomi movement, social protest, 153–157
organ donation, donor reluctance tied to
    unwarranted fears, 85
organic agriculture
    Nazi ideology, 59–61
    nitrogen use, 97–98
    pesticide denial, xvi
    weed elimination methods, 114

organic canned soups, salicylic acid,
    112–113
organic chemistry
    essential to twentieth-century
        agriculture advances, xvii
    mauve decades, 14–15
    origins of, 10–11
    population growth effects, xvi
    vitalism's death knell, 43–44
organic chickens, USDA rules, 114
organic foods, USDA labeling rules, 114
organic pesticides, known carcinogens, 91
organic produce, superiority debunked,
    110–111
ovulation rates, life security effect, 162

Parkinson's disease, rotenone, 91
PCR (Polymerase Chain Reaction), false
    positives, 120–121
penicillin
    antibacterial properties discovered,
        17–18
    discovery of, 17–18
    mass production methods, 19–20
peptide "zip codes", molecular targeting, 23
pesticides
    organic agriculture denial, xvi
    USDA approved for organic foods, 114
pharmaceuticals
    development of, 16
    peptide "zip codes", 23
phase contrast microscopes, gram stain
    development, 16
photosynthesis, efficiency improvements,
    130–131
physics, reductionism form, 21, 23–24
plant biotechnology, necessity brought on
    by population growth, 130
plant nutrition knowledge, food production
    contribution, 96–97
polyhydroxylated compounds, vital to all
    known life-forms, 6–8
Polymerase Chain Reaction (PCR), false
    positives, 120–121
population growth
    global middle-income group on the rise,
        128
    hunger/poverty rates, 128
    Industrial Revolution effect, 96
    life expectancy statistics, 146–148
    mass famine forecasts proved untrue,
        129
    need for new technology, 129
    per capita caloric intake, 127–128
    plant nutrition knowledge contribution,
        96–97
    statistics, 95
    synthetic fertilizer's role, 103–108

positional notation, mathematics
cornerstone, 37
precautionary principle, double standard,
88–90
precautionary principle, vitalist beliefs,
83–85
proteins
20-amino acid building blocks, 98–99
allergenic, 108–114
Bt (Bacillus thuringinesis), 109–110
name origins, 13
not a genetic information carrier, 29–30
RNA role, 32
thioredoxin, 109
purines, name origins, 14
pyrethrum, organic pesticide, 91
pyrimidines, name origins, 14

quantum mechanics, unification of
chemistry and atomic physics, 27

racial hygiene, Nazi support, 72–73
reductionism
advances in human health and well-
being, 24–25
described, 21–22
life-saving advances in medicine, 21–26
must demonstrate ability to generate and
explain complexity, 22
Ockham's razor, 4
problem stating as element of problem
solving, 24–25
scientific understanding of complex
systems, 23–24
"smart" pharmaceuticals, 23
variant interpretations, 22–23
vitalism as impediment to
understanding, 22
reductionist medicine, versus homeopathy,
58–59
reductionist science, nitrogen, 103
references, online searches, xii–xiii
regulator genes, DNA research role, 31–32
rejected knowledge
animal rights, 72–73
cosmic ice theory and Nazism,
76–78
DDT equated to chemical warfare,
60–61
disenchantment effects, 62–63
environmental deconstruction Nazi
style, 54–55
Holocaust as logical outcome of modern
science, 67–68
Holocaust compared to the practice of
medicine, 55–56
homeopathy versus reductionist
medicine, 58–59

ideas as responsibility of those who
hold them, 78–80
myth versus science, 65–66
Nazi comparisons, 53–63
Nazi ideologies and postmodernist
beliefs, 68–69
Nazi opposition to anthropocentrism, 71
Nazi opposition to humans dominating
nature, 66–67
Nazism and occult belief systems,
73–78
Nazism's landscape beliefs, 70–71
racial hygiene, 72–73
sense of purity belief dangers, 61–62
versus verifiable knowledge, 80–81
resources, unnatural, unfixed, and infinite,
xviii
ribonucleic acid. See RNA
rice, photosynthetic efficiency
improvements, 130–131
risk, from change, 159–160
riskless change, described, 151
RNA (ribonucleic acid), genetic
information flow, 31–32
romanticism, as reaction against science's
presumed dangers of knowledge, 3
romantics
belief that one can grow plants without
nutrients, 134–135
Chipko (tree huggers) movement,
136–137
deep plowing practices, 135
lack of understanding of the laws of
conservation of matter, 133
living foods diet beliefs, 143–144
natural food is not necessarily better,
138–144
selective data use, 133
social policy, 148–149
uncooked food passion, 142–143
water prudent food crop belief,
134–135
rotenone, organic pesticide, 91

salicylates, food sensitivity cause, 113
salicylic acid, organic canned soups,
112–113
science and technology
better world opportunity, xi–xii
cell discovery, 12–13
confusion with myth, 65–66
demystification of, 45–48
gap bridging benefits, 27
green revolution successes, xvi–xvii
human machine's origins, 3–5
humanistic values, ix
life improvements outweigh harmful
effects, viii–ix

science and technology (*continued*)
   "local ways of knowing" myth, 34–38
   Nazi comparisons, 53–63
   no need for vitalist principles, 34–39
   ongoing inquiry, 124–126
   organic chemistry recognition, 10–11
   origins of life, 6–8
   precautionary principle, 83–85, 88–90
   pursuit of truth, xv–xvi
   refutation of spontaneous generation, 42
   resource creator, xviii
   romanticism's reaction against, 3
   testable knowledge, 9
   transcultural force, vii–viii
   verifiable knowledge, 80–81
   virtue of incremental change, x–xi
   world population growth challenge met,
       129–130
scientific nutrition, human machine's
   origins, 3–5
second law of thermodynamics, entropy is
   increasing in the universe, 4–5
sense of purity, belief dangers, 61–62
Shell Petroleum Development Company,
   157
social protest, NGOs, 153–158
sources, online searches, xii–xiii
soybeans, ATF (anti-tryspin factor), 142
spectrum, visible light, 163
spontaneous generation, refutation of, 42
statistical mechanics, unification of
   thermodynamics and mechanics, 27
stem rust, Hope gene as limiter, 115
streptomycin, development history, 19
structural genes, DNA research role, 31–32
sulfa drugs, development history, 16
synthetic nitrogen fertilizer, population
   growth role, 103–108

T gene, barley, 115
thermodynamics, entropy crisis, 4–5
thioredoxin, non-allergenic protein, 109
thymine (T), DNA chemical unit, 99–100
transgenes, appearance in Mexican maize
   landraces, 120–124
transgenic food
   fear compared to nuclear power, 84–85
   Internet Frankenfears, 85–87

UNDP (United Nations Development
   Programme), 118
United Nations Development Programme
   (UNDP), 118
uracil (U), RNA molecules, 99
USDA, organic food labeling rules, 114

vaccines, pros versus cons, 152
vegetarianism, Nazi ideology, 59–61
vegetarians
   biodiversity harm, 107
   misguided reasons for reduced meat
       consumption, 106–107
verifiable knowledge, versus rejected
   knowledge, 80–81
victimless change, described, 151
victims of change, identifying, 153
virtue, vitalist beliefs, 90–92
vitalamines (vitamins), discovery of, 17
vitalist beliefs
   anti-genetically modified food mania,
       xvi–xvii
   conservation of wildlife over human
       life, 92–93
   development history, 41–42
   everything was endowed with
       "polarity", 41
   ferments, 13
   genetic inheritance, 28
   harmful impact, vii–ix
   homeopathic medicine, 45–48
   homeopathy versus reductionist
       medicine, 58–59
   interpreting the world, 34
   "local ways of knowing" superiority,
       34–38
   morality relationships, 42–43
   natural is safe until proved harmful, 146
   opposition to synthetic fertilizers,
       49–50
   organic molecules form only from other
       organic molecules 10
   precautionary principle, 83–85,
       88–90
   pseudoscience currents, 41–42
   purity of the past, 56–58
   re-enchantment of science, 45–48
   rejected from modern scientific
       knowledge, xvii–xviii
   rejected knowledge, xi, xv–xvi
   rejection of synthesization of organic
       compounds, 41
   religious overtones, 42–43
   scientific understanding is unacceptable,
       43–44
   spontaneous generation, 42
   uncooked food passion, 142–143
   uniqueness of life, 10
   verification incapability, 9–10
   virtue, 90–92
vitamin C, reduced by cooking, 143
vitamins (vitalamines), discovery of, 17

Waldorf School, holistic teaching centre,
    51–52
wildlife conservation, precedence over
    human life, 92–93

X-ray observatories, 166–167
X-rays, astronomy, 166

Zapatista Army of National Liberation
    (Mexico), 157
zero notation, mathematics cornerstone, 37
zero risk, defenses against, 152